CONTENTS

DEDICATION

Much love to my wife, Carmen; son, Dave; daughter, Devan; and grandson, Aden.

Encouraging the men and women in the business agility community to continue with shifting the ways of working to be more people focused and anchored on delivering value to delight customers.

ACKNOWLEDGEMENTS

Special thanks to Diana Larsen, author of *Agile Retrospective* and co-founder of Agile Fluency Model; Marty Nelson, Alchemy Code Lab founder; Dave West, CEO of Scrum.org; Howard Sublett, former CEO of Scrum Alliance; and Mike Cohn, founder of Mountain Goat Software, for lending their voices through our interview sessions to describe the meaning of delivering value.

Allison Pollard, our conversations about agility are endless, and we can go on forever. I sincerely enjoy our topics and look forward to the continued discovery.

FOREWORDS

Diana Larsen

My first impressions of Dr. Dave Cornelius are forever linked to his passionate commitment to the 5 Saturdays STEAM program. He created a program to introduce young people to the excitement of programming and technology. It expanded their minds and future of work possibilities for the 21st century.

Over later years, I learned how that devotion to value went beyond the young people in his programs. It extended to all the diverse communities of people involved in software development. It went even further to consider the value in organizations those people serve. I saw Dave's core drive in action. It lay in finding new ways to produce and deliver value.

As long as I've known him, his vision has been BIG. That big vision has grown to the scope of this book you hold.

In it, Dave describes the role of agile software development (in its many forms). He adds his awareness of the need for business agility in today's fast-paced world. Our awareness of the complexities we face gives us discomfort. It means change is coming. The pace of technological change fosters social and personal anxieties for many.

Dave reminds us in this book that our challenges are knowable and surmountable. We must adopt a spirit of discovery and learning. We keep in mind the value that emerges when we enable people and processes to do good. Like the 5 Saturdays STEAM program helped our children find their place in this challenging world ... we can find our place too.

Deliver Value helps adults focus on creating the valuable outcomes we need and want. In it, Dave guides us toward empathy and generative agile leadership. They lie along a path to better outcomes. He shares useful models along the way to prompt our thinking and action.

Dr. Cornelius honored me with a request to participate in the interviews for this book, and I'm glad to find my place in it. As you read it, you will find yours too.

Diana Larsen

Coach, mentor, consultant

Author, *Lead without Blame: Building Resilient Learning Teams* (2022)

Co-creator, "The Agile Fluency Model: A Brief Guide to Success with Agile" (2018, online and eBook) and its associated materials

Zachary S. Brooks, PhD

Dr. Dave Cornelius sets the foundation of *Deliver Value* by highlighting how "value" is most clearly evaluated through the lenses of societal struggle and accomplishment. Events such as the election of Barack Obama as the president of the United States are to be seen as innovations as they open up value henceforth not seen.

Taking a renaissance view of the 21st century, Dr. Dave asks readers, like he asks himself: *Am I delivering value?*

Then he answers the question by defining value, weaving stories to highlight business practices that create value (or not), and finally setting the stage to recognize and share your best value. *Delivering Value* reads as a compelling drama with a focus on the ultimate human applied science: the practice of business and delivering value in all aspects of life.

The premise should make you choose the book. The content should please you. And the effect of reading the book will make you refer to *Delivering Value* for decades to come.

Zachary S. Brooks, PhD

Amazon #1 bestselling author in four categories and author of *Discovering Your Human Algorithm: How to Live with Meaning and Purpose*

Allison Pollard

I first met Dr. Dave Cornelius along with one hundred other new-to-me Agilists in 2015 at an Agile Coach Camp event in California. I was struck by the depth of his knowledge (a reflection of his doctoral research and continuous learning), breadth of experience (as an entrepreneur, consultant in Fortune 50 to Fortune 500 companies, and coach for nonprofit organizations), and most of all, his humility. Every interaction with Dr. Dave since then has contributed to making me a better leader, coach, and person.

Throughout my journey as a project manager to Scrum master to Agile coach, I have been working with leaders at all levels to deliver value for their customers. Too often, customers react to new features like my college roommate did when she reviewed my homework assignment: underwhelmed by what they've been given. And unlike my college experience, a lack of customer satisfaction with digital products is feeding into a pattern of overwork and longer cycle times in organizations. Leaders and coaches today are seeing and experiencing these effects as the great resignation takes its toll.

The progression of my consulting career has allowed me to see firsthand the shifting role of leadership as leaders are expected to do more with less. We are increasingly reminded by global events how connected we are to one another and how change can impact us at a deeply personal level. Being a leader or coach during the pandemic has meant confronting feelings of confusion, loss, and helplessness that can pop up or sneak in unexpectedly within ourselves and our teams. Dr. Dave's insights on how to navigate innovation in today's changing world and call for generative leadership are both pragmatic and inspiring.

Perhaps surprisingly, breaking out of this pattern and creating an environment where people thrive is within reach. As you read on, you may recognize yourself or parts of your organization's journey in the stories you find. All of it is based on real people, real organizations, and the real problems that come up. They did not have all the answers to make their workplaces better, and you don't need to either. Borrowing the words of Ashanti, an executive you'll read about later, you can set an invitation and be the partner for teams to "practice agility to be able to weather the storms and enjoy the sunshine of an uncertain future of work."

To say that Dr. Dave's book is timely is an understatement. This book reminds us of the real, practical, positive value that we can provide for our team members and our customers in this world ... and for ourselves. Now is the right time to show up differently as leaders and create environments for people to thrive.

Journey on,

Allison Pollard, director of Agile Coaching, Improving Inc.

PREFACE

I would like to begin by asking you to pause for a moment, take about 15 seconds, and say out loud, "I am valuable." Next, call, text, or go on social media and say to one person, "You are valuable," using hashtag #Agile4Humanity. In this season of pandemics, social injustice, and economic challenges, we have to remember our value and the value that we create. People are capable of delivering value no matter their ethnicity, gender, religious belief, or other identifiers that place them in a specific group. If we value ourselves and other people, it is easier to see the value that we deliver in our family, our local and global community, and our place of work. The perception of value is real. We should not take for granted the goodness that we all have to offer. The ideas that spring from our minds and launch us into action to begin a journey of developing a product, service, or community have the potential of delivering value for our lives and the people that we convince to join us. It is amazing that we are able to build something from nothing.

I have been fortunate to be in the agile community and organization transformation business for the past twenty years. The last thirteen years have been filled with organizational transformation while working as an organizational coach and trainer in agile practices, lean startup, design thinking, and improving business capabilities through digitization to deliver value faster. Adopting agile values and principles into my personal life allowed me to introduce the concepts to my son and wife. In fact, everyone I meet receives some nugget of information about agility.

My career has taken me into the lands of leading as an employee and consultant in for-profit and non-profit companies globally. Sometimes the question that comes to mind is "Am I delivering value?" The "am I delivering value?" question is one that I frequently hear from peers and clients as well. Ok, perhaps most of the time, I question the value delivered by my products, services, and skills. Can the value that is being delivered be measured in a meaningful way?

Being an entrepreneur is a risky business, but being a black entrepreneur is even riskier. I recall the launch of my first business. The advice received from both black and white leaders was that I had to have a white person as the CEO of my company because people would not perceive my value the same way. At first, I questioned that white supremacy mindset because I grew up on the island of St. Croix, where kids were taught that we were capable of delivering value. Eventually, I gave in and believed what they were saying at the time based on my observation of the leaders of technology companies. I was living in Chicago and working in the software and technology industry in the 1990s. It was a time when technology was still peeking out of the shell to explode into the exponential growth that we have experienced. The internet was born as a commerce marketplace to allow people to buy and sell goods. Reflecting on that era, I remember the bulletin boards and modems that we used to connect with people throughout the United States and the world.

In 2008, my perception of who could lead a technology company changed. The first African American president was elected. Barack Hussein Obama became a symbol of possibilities for young and old black men in the United States. People may not realize the value that was delivered in that historic moment. My mother said to me that she did not believe that it was possible to have an African American president in her lifetime. It was a sentiment that I also accepted as being true. That was a common message that I heard in the African American community throughout the United States and beyond. My socio-economic status did not change the perceived reality of the constraints that have been placed on non-white people. The election of Barack Hussein Obama as president of the United States was truly a breakthrough innovation.

The sense of being fully able to achieve anything returned, and I felt the familiar emotion of wonder from growing up on the small island of St. Croix, USVI. The disparities found in the science, technology, engineering, and math (STEM) industries by gender, race, and socio-economic levels were created by the perception that these groups of people were unable to deliver value. The stereotypes of women being too emotional and black people lacking the mental fortitude to innovate in STEM fields are the result of devaluing human beings to match a four-hundred-year-old narrative globally. Unfortunately, that narrative is still alive and growing. People who are devalued must create their own narrative of value. They must not stand still and cry out, "Woe is me." We are capable of delivering value and do not need permission to be successful. Fortune favored me in 2015, when I received a letter from President Obama for the volunteer work performed in under-estimated communities by teaching science, technology, engineering, agile, and math (STEAM) to high school students. The A in STEAM usually describes art, but for our adaptive purpose, the A stands for agile practice. The program that I founded is called 5 Saturdays (www.5Saturdays.org). I am grateful that my non-profit organization is still able to share industry knowledge in STEAM in the United States and beyond.

The year 2020 was the year of turmoil, pivot, and transformation for people throughout the world. A pandemic ravaged the world and did not care about the false narrative—the only value sought was a live host for the COVID-19 virus to feed on and live. Millions of people lost their lives because of the COVID-19 pandemic. The social justice unrest grew because of the killing of unarmed black men in the United States, and the Black Lives Matter movement sprung wings. An awakening took place that included people from various segmented groups realizing that black men were devalued by primarily white police officers and society—taking unarmed black lives was not punishable by a system that should protect all of its citizens.

My disclaimer is not all police officers are bad or commit the ultimate action of taking someone else's life. I can recall my positive experiences with police officer Renholdt "Rookie" Jackson growing up. He ran the Police Athletic League (PAL) club that benefited young black boys in my community on St. Croix, USVI. He provided positive encouragement and opportunities for many of us. When black boys were labeled as bad and not worthy, he found a way to create space for conversations and uplifting. Officer Jackson provided a tremendous amount of value that still resonates with me today.

The world's economy was affected by small businesses closing their doors and large corporations laying off employees. Many industries disappeared because of the shutdown of our activities to reduce the spread of the COVID-19 virus and deaths. Small business owners became broken hearted and disillusioned as their dreams turned into nightmares.

The rise of president 45 and his authoritarian leadership began to chip away at the norms of democracy in the United States. We are as divided as we ever were. Again, in my lifetime,

I could not imagine that the leader of the free world would use every crooked method available to nullify an election. We have a divided country, and 70% of the seventy-four million people that voted for president 45 believed that he won re-election. The value that president 45 wanted was unlimited power and to enrich himself and his family.

Most people would like to view 2020 in their rearview mirrors, nothing more than a distant memory. So much has happened in 2020 and 2021 that speaks to our resilience, loss, and agility. We can adapt to changes, which has been proven with our endurance to the new societal norms. We need to sit and ponder: how do we go forward and deliver value in the new norms of working and living?

This book is focused on how we navigate innovation in a volatile, uncertain, complex, and ambiguous (VUCA) world. You will experience fictional short stories to help tell the stories about people working to create value. The fictional stories are centered around the character of Ashanti MWendo, a new chief product owner (CPO) responsible for changing the culture in the organization to deliver value. Real stories are provided to help us connect with our own journey of delivering value. The goal is to introduce characters that you may connect with as you read or listen to the journey of delivering value.

INTRODUCTION

People working to enable an organization to thrive in a volatile, uncertain, complex, and ambiguous (VUCA) world that is accelerated by the Fourth Industrial Revolution (4IR) must find multiple ways to deliver value. The COVID-19 global pandemic is another accelerant that ignited digitization of business operations and remote working environments. During digital transformation efforts, workforce capability building is vital at every level of the organization. The focus must be to enable happy contributing people, satisfied customers, and thriving business activities that bring clarity through measurable outcomes. When leaders are asked how they know their hypothesis about a new initiative will deliver the value promised, the answer should be informed through measures that are supported by primary (people) and secondary (research platforms) sources. The practice of vanity measures to satisfy each leader's ego and position has become irrelevant. Vanity measurements are easy to measure but rarely describe the impact to the business operations (Forsgren, Humble, and Kim, 2018).

The mindset for leaders and team members must shift to deliver a sustaining value that is resilient. Being competitive in a global market will require the adherence to simple principles that encourage innovation and continual learning to build sustaining value for their people, customers, partners, and community. The effort is not only for assigned leaders (C-Suite, VPs, directors, and managers) but also for the emergent leaders who are courageous enough to step into opportunities to lead and provide creative and critical insights.

Developing workforce capability is vital to combat the phenomenon of the great resignation of 2020 to 2021. The great resignation phenomenon is people resigning from their current jobs for other opportunities and retirement. This was particularly high in the technology and health care industries. One way to mitigate the great resignation phenomenon and the effect of the 4IR is through reskilling existing people in the organization. The reskilling efforts within the organization and the promise for new talent to join the organization will lead people to an abundant mindset to develop a work environment that supports new opportunities, capabilities, purpose, and autonomy. Leaders must develop a strategy to support happy contributing people in order to retain resilient people.

The status quo approach to delivering value will not work in this new economy because customers have choices and market competition is rising globally. Customers have the choices to hire and fire services and products given the digitization of most products and services with access via the internet. Effective responsiveness to customers' needs allows customers to feel valued. Obtaining insights into your customers' sentiments is equally vital to being able to respond appropriately.

As a professional coach, I am an aspiring creator of value, and my mantra is "helping people achieve awesomeness." This is a tall order to achieve but a worthy journey in all aspects of life. Helping people achieve their level of awesomeness is a partnership that I engage in with people that I coach.

However, the ultimate ownership resides with the individual(s) working toward a *meaning* of awesomeness. When we say someone is awesome, we are describing the meaning of value at the highest level. The words that come to mind are awe-inspiring, magnificent, amazing, impressive, and wonderful. These words are synonyms for the word awesome.

The "pursuit of happiness" was written into the 1776 United States Declaration of Independence. The "pursuit of happiness" is an American dream, but not exclusively. I want to assert that happiness is a state of satisfaction. When people achieve their level of awesomeness, it is one state of satisfaction. As we deliver value, the state of satisfaction is being given and received.

Happy contributing people and satisfied customers are the first and second goals required to enable a thriving business. Thriving business is the ultimate goal that applies disciplines that support easy and nimble adaptation of changes in a competitive landscape that will elevate a failing business to a thriving business.

DEFINITION OF VALUE (DOV)

If you ask several people what the definition of value is, you will likely receive varying responses. Let's see what is written about the definition of value in the *Merriam-Webster Dictionary*. This dictionary was established in 1828. Value is identified as a noun, verb, and adjective. The following are the three definitions as a noun that I selected:

"1. The monetary worth of something: market price
2. A fair return or equivalent in goods, services, or money for something exchanged
3. Relative worth, utility, or importance"

Value is the experience of receiving something useful that is beneficial to us. Value is also giving something of merit to others that is good or helpful. The recipients of value are customers as well as people at work, in our community, and in our family. We want to receive value when we engage in a transaction with a company or interact with people. But value has a different meaning to each person, group, or company. This is because we have different experiences and expectations based on background and culture. We all desire to receive value. My definition of *value* is outcomes that are measurable, realizable, and shareable.

I spoke to four people who have an influential opinion in business and the agile community to discover their definition of value. Each person interviewed provided a unique context based on their own experiences.

Definition of Value (DoV) - Influential Leaders' Voices

Marty Nelson, Alchemy Code Lab founder:

Value has a positive exchange, positive emotion, and experience. Value is generative— money, friends, and social connections, working ecosystems. Value is something that satisfies true wants.

Diana Larsen, author of Agile Retrospective and co-founder of Agile Fluency Model:

I think there are different kinds of values. It's too bad we only have one word to talk about several definitions.

One context is value to the customer. Businesses are sensitive to their customers' needs. They want to give the customer what they want so that customers will pay for services and products. So, there's the value around the value stream going to the customer. Are we getting closer and closer as time goes by to building what the customer wants and will accept and pay for? Product management helps the organization focus on the items with the highest priority over others, which gives a sense of, "This is more valuable than that, at least at this moment in time."

To better understand priority, we can add methods like the cost of delay, and it begins to get us closer to understanding value as well.

Then there's also business value, which is the investments enabling the business to thrive, be resilient, and sustain over time. Value, I think in the business sense, is very much tied to strategic direction. Is this strategic position providing us value because it's moving us more toward our strategic goals, like targets in the *Improvement Kata*? Is this moving us in the direction that we need to go to give greater value to the business? In the Agile Fluency model community, we tend to talk about value as business benefits. The value is the benefits that the business wants to gain from supporting and investing in their teams to help achieve the maximum value from the products in the market.

Dave West, CEO, Scrum.org:

Value is a subjective measure. Therefore, it is not how I define it that matters, but ultimately the person and organization that I wish to bring value to that are relevant.

For example, it would be easy to describe the outcome of a sale (happy customer, rich seller, etc.), but ultimately, you must ask why is the person buying that thing—what is the value they want to receive? Also, what value does the seller achieve—why are they selling this thing rather than doing something else? It is a complex question—so ultimately, I say the value is something defined by the individual that describes the outcome of the event.

Howard Sublett, former CEO, Scrum Alliance:

In my mind, value is a relationship between the amount someone pays and what they *perceive* that they get in exchange or the amount of effort that's expended for the perceived return. It can also be the amount of joy you perceive in return.

For example, if I were to drive across the country to visit the Grand Canyon in Arizona from New York City—perhaps one of the things on some people's bucket lists. If I invest a lot of time, money, and effort to get there and when I arrive, the weather is foggy and rainy. I can't see the views or experience the canyon in the way I'd envisioned. Then I will *perceive* the experience as a low value even though it's generally considered a high-value thing to do.

Alternatively, a low-effort, high-value experience for me is sitting on the back patio and having coffee with my bride. We don't get to do it often with me traveling so much lately, but when we do, it requires minimal effort and returns an incredible amount of value to me in my personal development. So, value is the relationship between what you invest or pay and the amount of joy you perceive that you received in return.

Outcome Values Are Measurable

My definition of *value* is outcomes that are measurable, realizable, and shareable. If you can measure it, the possibility to realize the outcome increases, making the outcome shareable with others in the organization and even with partners.

Outcomes are the payoff or benefits received from a specific interaction or transaction. The measurable outcomes can be quantifiable or qualifiable, demonstrating that the benefits are validated learning and not a hypothesis. Objectives and Key Results (OKR) is one method to measure the outcomes. Google, rock star Bono, and The Gates Foundation used the OKR method to identify what was the most important for the next three, six, or twelve months (Doerr and Page, 2018). The OKR structure is broken into two parts that include 1) an objective (why is this important?) and 2) key results (benchmark and monitor how to get to the objectives).

An OKR example to demonstrate measurable value follows:

Objective: Our virtual reality immersive learning course improves food safety.

Key Results 1: 85% of the people involved with production/ handling of foods demonstrate a high commitment to our food safety standards.

Outcome Values Are Realizable

The second value attribute is realizable, which represents the actualization of a desired or even an unknown outcome. Maslow (2013) described the five levels of human needs as five steps in a staircase as follows:

1. The first step is physiological (food, water, warmth, and rest).
2. The second step is safety (security).
3. The third step is belongingness and love (intimate relationships and friends).
4. The fourth step is esteem (prestige and feelings of accomplishment).
5. The fifth step is self-actualization (achieving your full potential, including creativity).

The realization of the measurable outcome is the fifth step in the staircase of actualization. It is not driven only by one's capability but also fueled by collaboration with other people. Arriving at the desired outcome is rewarding. Remember a time when you worked really hard to produce an outcome that benefited your client or organization? Realizing the outcome was worthy of celebrating the moment with your team members.

An OKR example to demonstrate realizable value follows:

Objective: Our virtual reality immersive learning course changes food safety standards in the U.S.

Key Results 1: The Food and Drug Administration (FDA) adopted our course as a standard by Q2 2023.

Outcome Values Are Sharable

People are seeking to obtain value in many areas of life. It is the reason that we spend a tremendous amount of time on social media sharing lived experiences or desires. We are often on a quest to find value through our experiences. The concept of value as an experience or outcome to be shared provides a way for others to validate the perception of value.

An OKR example to demonstrate sharable value follows:

Objective: Our virtual reality immersive learning course changes food safety standards globally.

Key Results 1: 85% of the companies in the food industry learned about our food hazards and risks simulation module from their employees.

Business Value

The next lens is the value from a business perspective. Members of most business organizations often pursue maximizing value through effectiveness and efficiency. Optimizing flow is an ongoing effort and part of continual learning. We check to verify the satisfaction of customers, business leaders, and teams with an optimized flow state and decide to acknowledge that it is good enough or identify if more changes are needed.

Identifying the need for change is not enough because an associated action is required to ensure that we remain in an optimized flow state. Deming's (1950) continuous improvement model of plan, do, check, and act (PDCA) is an optimizing flow. The "check" state in the PDCA model relies on measuring the results and sharing that information with the appropriate stakeholders. Stakeholders are people who may have an interest in a specific result. The "act" in the PDCA is based on a decision to improve the current state or situation. To act or not to act depends on the results from the "check" metrics.

We should not make changes unless that data informs us of the action to take. Otherwise, we are reacting to gut feelings. The world of agility adapted the PDCA model and described the behavior as plan, do, inspect, and adapt (PDIA). The core of lean thinking is optimizing flow to minimize waste and achieve the greatest value. I want to focus on flow to continually move value from one person or organization to another with limited friction. Some would like to think that interactions can be frictionless, but that is an ideal state of being. Friction exists in everything we do, but we can minimize the amount of friction to make it appear frictionless.

Customer Value

I invite you to pause for a moment during your next visit to a coffee shop and ask: what value is provided to the customers and employees?

The coffee shop provides a customer experience that makes people feel accepted during in-store or online buying experiences of a coffee drink and a light snack. The cozy community feel can help people connect and improve the exchange of value between the barista and the person waiting for their drink before going on their way to work or school. The internet cafe, family-style, and coworking collaboration desk configuration make it possible for students and business professionals to connect for a conversation or do homework.

The people working at the coffee shop tend to have a positive attitude toward customers, and people buy from people they trust. The coffee shop employee receives many benefits that help them finish college or establish an alternative career. The coffee shop also promotes social impact and change in communities to feed the homeless through unsold foods.

My Career Delivering Customer Value

One of my claims to fame is that I delivered more than $1 billion in customer value over my career—through the software platform that I led in my role as a senior manager of product development.

The calculated value was based on the number of transactions and dollar value pushed through the software platform I was responsible for developing and managing. Around 2004, the financial markets were thriving, especially the housing industry. People and companies were making money hand over fist, and accelerated innovations were needed to enable faster transaction throughput. People do not scale. The only way to achieve sustainable faster throughput is by leveraging advanced technology platforms. Rarely do you get to grow the value of a technology platform in a company where the core business is not delivering technology as a value proposition.

I inherited a technology platform that did not scale and was brittle. The systems were responsible for collecting credit ratings, home appraisals, and title information for home buyers, brokers, and other financial organizations. It was estimated that $80 billion in transactions per month were being processed across the financial organization. The company revenue was four basis points or 0.04% on the $80 billion total transactions. That is approximately $320 million per month in revenue opportunities. A delay in processing those transactions affected many people, including customers, salespeople waiting for commissions, and the organization's revenue. The delays created an immense number of tensions between the technology group and the business operations group.

At month-end closing, all hands were on deck to watch the system twenty-four hours a day for seven days. We did not get much sleep, and the wonderful Blackberry mobile device that we carried would sound off at two a.m. or all hours in the morning, alerting us that the system was broken. It was even worse during the peak hours of the day. I asked my boss one day how he was doing, and he said that he felt like he was being abused in jail.

That was a powerful statement, and I tried to empathize with my boss's experiences daily. The VP of operations would publicly berate my boss and make him feel inadequate and inept at best. One day, I walked into the VP of operations' office and asked what the opportunity cost to the organization was when the system's performance was slow or unavailable. He looked at me with a surprised facial gesture. Still, I forged forward to let him know that as a product manager, it was important to understand the value of my assigned products to internal and external customers. During peak week, he and I spent several days collecting information about transactions stuck in the pipeline to understand the ripple effect financially on internal and external stakeholders. It was important to know how much money was not being processed.

Finance departments frequently leverage the "float" to make money for their companies. For example, if a company could process $100 million and collect a day's worth of interest, that can be a significant annual revenue opportunity. This was a worthy story, and stakeholders would appreciate the value IT enabled through the solution platform. The knowledge and understanding of the impact of poor-performing technology platforms also resonated with the software development teams, and they began to look for solutions that could help stabilize the brittle systems.

Our team collaborated to fix the performance issues and provided increments of value to the internal and external customers within three months. This was a great step toward reducing uncertainty and enabling the organization to realize value without delays. The cost of delay is an important concept for business and technology people to understand. Having the ability to realize the value and use it immediately was key. My boss and VP of operations began to work better because of the improved system stability. Value was realized in respect and collaboration between the two leaders. My boss did not have to be on the defensive, and the VP of operations reduced his antagonistic behaviors about the system performance.

Organization Leaders Want Value-Driven Outcomes

Tell me how you will measure me and I will tell you how I will behave.

— Eli Goldblatt

The conversations that I commonly had with organization leaders can be summarized into the desire to maintain or increase gross margin, increase brand loyalty, attract or retain resilient people, practice agility, and leverage risks for exponential gains. These are the five core outcomes that leaders are often measured against. Gross margin (GM) is the amount of money earned from the sales of products or services.

A higher GM value indicates the profitability of a company, which means more money is available for investments. It's a sign of financial health. To calculate GM, you subtract the cost of goods sold (COGS) from net sales revenue.

Leaders become concerned and frustrated when GM is negative because it is a sign that expenses or sales are not performing as planned. The response is often the proverbial *tightening of the belt* that leads to layoffs and cost cutting. However, when GM is positive, leaders can use this information to drive strategic decisions that include offering discounts to customers to increase sales and outpace the competition. GM should not be the only key performance indicator (KPI) that leaders should use to determine profitability because not all expenses are included in the calculation.

The importance of the pursuit to protect and expand brand loyalty creates an environment that will help to create happy customers and a healthy brand, which often leads to recurring revenue streams. My great-grandmother would often tell me, "A bird in hand is better than two in the bush." The meaning is to focus on the existing customers and not only pursue new customers. The cost to attract new customers can be significantly reduced when a net promoter score (NPS) is high and people are excitedly sharing about a product and service to people in their network. NPS is more than just a score; it represents actual feelings about your products and services. Buying a new or expensive product is an emotional activity, and it creates anxiety about whether the product or service will provide the desired value or meet expectations. Brand loyalty is not limited to products and services but also applies to individuals delivering value to people and organizations. Creating brand loyalty as a coach and trainer is as important as Apple's or Amazon's brand loyalty.

Employee NPS (eNPS) is another significant detail that is important to leaders who want people who are fanatically loyal to the organization and willing to provide innovative ideas to expand the company's growth. If employees are happy and productive, the likelihood of customers being satisfied is significantly greater.

Emergent and assigned resilient leaders help to steer the organization and create an environment where people can thrive. The ability of an organization to attract and retain resilient leaders creates continuity of ideas, culture, and knowledge. The intrinsic values of the resilient leader include the following:

- Enable flow
- Ability to get up when times are tough
- Find a way during impossible odds
- Supportive regardless of the circumstances

Resilient leaders are those steely nerved men and women who are capable of discovering ways to push through to take advantage of opportunities and innovate to reach new markets. Oprah Winfrey is top in my mind when it comes to women who are resilient and transformative in their fields. Germany's first female chancellor, Angela Merkel, is a top leader in Europe and has moved her country out of the financial crisis into a healthy financial status. The late William (Bill) Campbell, a former football coach and entrepreneur, became a beloved and effective executive coach to leaders at Apple, Intuit, eBay, Yahoo, Twitter, and Facebook. He provided resilient leadership coaching and advice to many Silicon Valley leaders, which changed the trajectory of their companies. He was considered the "trillion dollar" coach because of the number of leaders that he advised and coached at multi-billion dollar companies in Silicon Valley. Bill was also known for his infectious bear hugs and air kisses that he used to let everyone know that he loved them in the sincerest way.

An illustration of leaders seeking value-driven outcomes can be gleaned from comparing leaders in the financial markets and Southwest Airlines leaders. The financial markets in 2008 went upside down because too many unfavorable risks were taken by financial companies, and the world went into a tailspin, resulting in many people losing a tremendous amount of wealth. Southwest Airlines, on the other hand, hedged long-term oil contracts to mitigate the turbulent oil prices that affect fuel prices. The practice has enabled the airline of "love" to maintain a healthy balance sheet and protect against the fluctuations of airline fuel.

The five desired outcomes leaders seek are 1) maximize value by being financially aware, 2) people autonomy, 3) extreme customer focus, 4) being nimble, and 5) taking the right amount of risks. The five growth areas allow leaders to evaluate additional strategic options to determine what directions to pursue. If things are going well, then the decision can be to stay the course; otherwise, pivot to achieve the right beneficial outcome for the organization.

Future of Work

Automation and artificial intelligence (AI) will play an important part in the future of work, but there will still be a need for people to create the algorithms that enable the AI to be effective. Most important is keeping the AI in check to minimize built-in bias for underrepresented people. People are the essence of the future of work, but their capabilities will be greatly enhanced by automation and AI.

The majority of innovations available to date were created by people. They produced something from nothing and made it available for the benefit of other people, the earth, and other inhabitants of our shared space. Computers were created by people seeking to deliver value faster by processing information at a rate that humans are unable to achieve. Cars gave us the ability to go faster and further than we could imagine, but people are the consumers and innovators of the new way of mobility.

The future of work is about people working together as a team or co-collaborators to produce an output or outcome. People are often dissatisfied with how they are treated, the opportunity to contribute, or the ability to perform at a high level. The recent phenomena of the great resignation have been the result of people pursuing opportunities that make them happy. There is a way to enable people to find purpose in their work and participate as members of a high performing team that are happy contributing people. I am using the phrase "happy contributing people" to characterize the feelings that people experience and the ability to contribute to value for customers, organizations, and themselves. We must not lose sight of the fact that people deliver value when they are in an environment that nurtures the ability to grow capabilities. Some people may not be able to achieve the happy contributing people state of being because of constraints in the work environment created by leaders and coworkers. Most of us have experienced the limitations imposed by others who are not willing to co-create an environment that gives us the ability to be happy contributing people. We may have persevered despite the imposed barriers, but how much better things would be if that was not the case.

Limiting beliefs may also be an impediment in people's ability to achieve the happy contributing people state of being.

Limiting beliefs are the mindsets that can cause you to say, "I will never be successful," "I cannot develop computer skills," or some other negative thoughts and beliefs that limit your ability to participate successfully in the experience of the happy contributing people state of being.

Almquist, Senior, and Bloch (2016) defined "four value elements: functional, emotional, life-changing, and social impact." The example of the financial company that I worked for was an illustration of business operations realizing predictability and stability through the technology platform that included four elements of value: functional, emotional, life-changing, and social impact. Functionally, the people working in the business operation centers were able to process loans without delays. Emotionally, the leaders found harmony after many months of dysfunction. The customers experienced a life-changing moment because they were able to realize a dream of owning a home. The social impact was families were able to move into communities, creating a stable living situation.

Happy Contributing People

Happiness consists in achieving, through the course of a whole lifetime, all the goods—health, wealth, knowledge, friends, etc. — that lead to the perfection of human nature and to the enrichment of human life.

— Aristotle

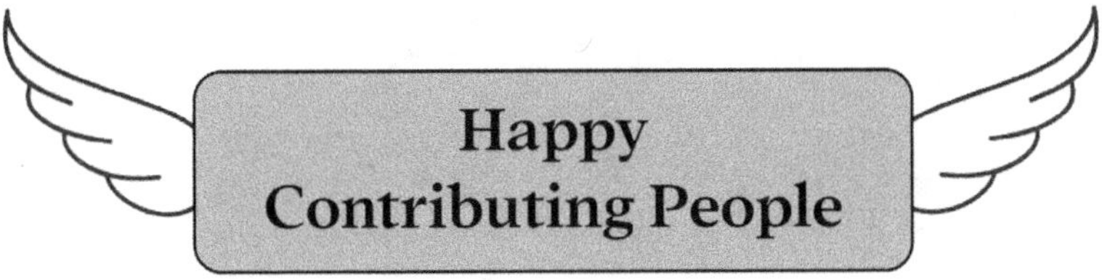

INTRODUCTION

Happiness is based on experiences that are metabolized in our mind, body, and spirit. When people say, "Think happy thoughts," they are referring to past experiences that we can recall and dream about. Oswald, Proto, and Sgroi (2015) found happiness made people around 12% more productive while unhappy workers proved 10% less productive. Positive emotions appear to invigorate human beings. People are the center of delivering value to customers, and they provide innovations that allow the organization to thrive. The more happiness people experience, the better opportunity the organization has to have more satisfied customers and to thrive.

Pink (2009) said if people were paid the right amount of money and given autonomy, purpose, and mastery, they would be motivated to be their best selves at work. Psychological safety is another popular topic that resonates with people. In a psychologically safe workplace, people experience the freedom and courage to express ideas so the individuals and organization can thrive. The same concept applies to enabling satisfied customers to volunteer personal time to show the world how amazing your organization's people are via social media or word of mouth. If happiness is such a critical emotion and experience to invigorate people, why do organizations invest so little to create an environment where happiness can be experienced?

LEADERS WANT RESILIENT PEOPLE

Resilient people are able to respond to challenges and find opportunities when it appears there are none to be found. They are the "pick yourself up and brush yourself off and keep on ticking" kind of person. Oprah Winfrey's story comes to mind as she was not a likely person to be a television personality. She was the daughter of a teenager and was herself a teenage mother at 14 of a son who died (Adams, 2016). Oprah is now one of the most successful entrepreneurs in the world.

Colonel Harland Sanders is another person with a resilient story. He broke through to success at the age of 62 with a $105 Social Security check in hand, pitching his chicken recipe to restaurants (Adams, 2016). According to reports, some 1,009 people rejected him before one accepted his offer. Kentucky Fried Chicken (KFC) is a national brand in the United States and beyond. Resilient people have grit and toughness to endure unfavorable experiences and rise above to achieve a desired goal.

David Martinez, principal at Early College High School and colleague, used the word GRIT as an acronym. The initialism translated into G is Growth, R is Resilience, I is Intensity, and T is Tenacity. My sister Lateefah and I came from humble beginnings and independently lived by the mantra that failure was not an option. I adopted the "by any means necessary" phrase used by Malcom X to help rise above self-doubt and limiting beliefs.

Resilient people can show up and be vulnerable without fear of others being unsupportive. Brown (2015) defined vulnerability as "the willingness to show up and be seen with no guarantee of outcome—it is the only path to more love, belonging, creativity, and joy." The definition of love is patience and kindness, not only emotional behaviors. Becoming comfortable with being vulnerable helps the resilient person or team to live in the moment of failure and to inspect and adapt when things are not going as planned.

The Apollo 13 operating missions' team was credited with the safe return of the flight crew, and they earned the Presidential Medal of Freedom (Granath, 2015). The spacecraft's oxygen tanks exploded 205,000 miles above the Earth, and the operations crew put their collective minds together to create an innovative tool that removed carbon dioxide from the lunar module, thereby enabling the crew to breathe long enough to return home. We currently use the phrase "Houston, we've got a problem" sarcastically when there is an issue with initiatives that are not going well, but it originated from the crisis experienced by the Apollo 13 team.

Have you ever wondered what special gene a resilient person has? The seven attributes that I have observed in resilient people are:

1. Motivated - have an internal drive to achieve specific accomplishments
2. Purposeful - demonstrate resolve and a focus
3. Curious - inquire and experiment
4. Growth mindset - seek to increase knowledge and capabilities
5. Dreamer - have an imagination to see what is possible
6. Failure responsive - learn from mistakes and improve

7. Perseverance - work through difficulties to achieve the goal in mind

Building a team of resilient people with the seven described attributes usually increases the organization's ability to take on challenges and sustain continuity to achieve the organization's strategic goals. Resilient people create space for leaders to focus on strategic opportunities and reduce the cognitive load of having to always be the expert. High-performing teams are often built on people with resilient attributes and GRIT. The experience of working with resilient people gives a boost in creative energy and leads to frequent innovations. Resilient people have arrived through trials by fire.

The seven attributes described are well suited for organizations that require business continuity and innovative changes to coexist to remain competitive. Imagine an established organization that has an entrepreneurship spirit to pursue opportunities that are not part of the status quo. Resilient people with GRIT are the right fit for this type of culture. The seven resilient attributes allow team members to work on new ideas that are yet to be created and navigate the uncertainty of new innovations.

HAPPY CONTRIBUTING PEOPLE FICTIONAL SHORT STORY

As Ashanti MWendo glanced out at the sunset painting vivid colors outside of her study, she thought about how far she had come. Being an African American woman born in the United States Virgin Islands, she had migrated to the mainland to attend college, unsure of what great opportunities awaited her.

Yet she always knew she was destined for more. Her surname, MWendo, translated to "motion our journey" in English, while her first name meant "thank you." And she was thankful for the journey that brought her here.

As she sat in her snug, comfortable office, surrounded by plants and photos that reminded her of the family she was doing this for, she found herself invigorated with excitement. Tomorrow, she was going to be starting her new job as the chief product innovator at a progressive company called Avant-Garde Realities (AGR). She had done her due diligence and researched the leadership and the espoused culture and read many articles about how great it is to be an associate of Avant-Garde.

She was excited about the technology company that created futuristic virtual reality (VR) simulations that allowed people to imagine their future of work, love, and many other desires, though the most popular VR simulations were exotic vacations. Ashanti felt like she'd be doing something meaningful by bringing these experiences to other people.

Of course, Ashanti wanted to bring her best work to this new team. Her career journey had traversed many large corporations that gave her the skills to imagine new products and services and bring them to market. This opportunity was Ashanti's first role in the C-Suite as the chief product development officer, and she was proud of where she landed, but that wasn't enough for her.

She would lead a group of four hundred people with a mix of skills that included product management, software development, UX/UI design, and architects. And she always wanted to influence an organization's culture to help them become happy contributing people.

As the soft, warm light of the lamp on her desk fell upon her papers, Ashanti considered the approach she'd take to this position. She decided she was going to apply the generative agile leader (GAL) values to guide the people directly entrusted to her care and influence others. To help her maintain her focus, she wrote out a list of these values.

1. High cooperation – Focus on we
2. Messengers enabled – We trust you to achieve our goals
3. Risks are shared – Win and lose as a team
4. Bridging encouraged – Partner and have conversations to learn more
5. Failures lead to inquiry – Ask, "What did we learn?"
6. Novelty implemented – Run the experiment and evaluate the hypothesis!
7. Abundance mindset – We have capacity to share with others

The GAL values #1 through #6 are derived from the work of Ron Westrum's (2004) "A Typology of Organisational Cultures," which states that organization culture can be viewed as:

1. Pathological (power-oriented),
2. Bureaucratic (rule-oriented) or
3. Generative (performance-oriented).

GAL value #7 represents the mindset required to share knowledge and success with others in the organization. The GAL value system combines the continual learning and performance-oriented context to create a continual learning and high-performing organization. More specifically, it helps develop an environment that enables continual learning and high-performing teams.

Ashanti leaned back in her chair, tapping her fingers on her mahogany desk. She wanted to make sure to apply purpose and humility in developing teams that delivered value for customers, the organization, and themselves.

"How might we attract and retain resilient people?" she wondered out loud. "How might we practice agility?"

From the beginning, GAL value #1 (high cooperation) was included in her thinking. What can be expected in a generative culture? An environment where people learn quickly and innovate to meet the demands of customers and markets. People collaborate more effectively with a higher level of trust in all directions of the organizational structure (across, up, down). The petty personal issues are dampened to elevate the goal of the products and services that are needed to sustain the business competitiveness.

Generative cultures enable diversity, equity, inclusion, and belonging. Ashanti knew that generative leadership creates space for development and performance growth for the people working in the organization and community. The generative environment's generosity extends beyond the doors of the organization and into the local and global community. Developing a generative business culture is fundamental to frictionless operations and effectively collaborating with people to succeed in a highly competitive world.

Ashanti scheduled a two-hour meetup for the following Wednesday morning at nine to connect with her teams to share the idea of using the GAL values. The meeting would be a hybrid meeting with some people meeting face-to-face and others attending via a video conference platform. In preparation for the meeting, Ashanti created a virtual space for the team members to read through some of her thoughts and to begin writing questions prior to the meeting.

That Wednesday morning, Ashanti was excited as the sun warmed her skin and brightened her day when she left her house. It was the day of the team meetup, and Ashanti couldn't wait to begin the journey of partnering with her teams to co-create an awesome working environment. She made sure she was in the right headspace going into this, going over her notes and preparing her meeting. Then, she got situated in the well-lit room and addressed her new team members.

She began the meeting by doing a simple icebreaker to engage the team members for the first thirty minutes. It went over well, and Ashanti knew this was a team she was going to be proud to lead. Then, she dug into the heart of the meeting.

"What I am about to ask you is to be a partner in the journey to create a workspace that you would want your own children in the future to work in," Ashanti said, her voice booming with pride. "This is an invitation and not a mandate. We want people here willingly and excited to make a difference.

"To make sure that our teams remain strong, we need to hire and retain resilient people. Each of you will play an important role in selecting the people that you work with. And together, we're going to build amazing products and services for our customers. We are going to practice agility to be able to weather the storms and enjoy the sunshine of an uncertain future of work. As self-organizing teams, you will have the option to select what lean-agile framework works best for you."

The room erupted into applause and cheers and a few shouting, "Hell yeah!" Ashanti was invigorated by their enthusiasm.

"For now, there will be a limit on the number of lean-agile frameworks," Ashanti continued. "These are Kanban, Scrum, or the Scaled Agile framework (SAFe). For those of you who are not familiar with these three lean-agile frameworks, allow me to provide some background information."

Ashanti pulled up an easy-to-read bullet point list on the screen.

- Kanban is a practice created by Taiichi Ōno and was first used by Toyota lean manufacturing. It has been adopted for use in software development.
- Scrum is a practice created by Jeff Sutherland and Ken Schwaber, which is widely used in software development teams.
- SAFe was created by Dean Leffingwell and is used by software development teams that are between 50 to 125 people.

"The goal is for you to decide which one works best for the work you are doing," Ashanti explained. "A trainer and coach will be provided to help with the adoption of the selected lean-agile framework.

"Some of you are already in assigned leadership roles, but I want the rest of you to not be afraid to step into leading when opportunities arise. I strongly believe in emergent leadership. I am asking you to dream with me so that our shared goal is to be a continual learning and high-performing team.

"Our definition of value is a measurable outcome that can be realized and shared. We will measure our progress and focus on delivering value for customers. When we take on Big Hairy Audacious Goals, we will realize them through incremental deliverables.

"Most important for us is to share what we know with others in the organization through demonstrating an abundance mindset. This will not be easy, but the journey will reward you with growth and satisfaction. 'Do or do not; there is no try,' Yoda says. I look forward to our journey together."

Manny raised his hand vigorously, which Ashanti took as a good sign. That meant they were engaging with the materials, which was what she wanted most of all. She kept control of the meeting, though, and allowed her team to speak one at a time, making sure their voices were heard.

"Ashanti, will we be able to select the team we want to work on?" asked Tarasi, a young but incredibly intelligent team member.

"That was the same question I had," said Manny.

"Let's take it to the teams and see what they say," Ashanti replied.

A chant began in the room, "Select our teams, select our teams, select our teams."

"Well, it sounds like you got your answer," she said.

As her presentation concluded, the music volume started with a whisper and gradually increased to a louder level. People filed out toward the exits in a random fashion. Some hurried along while others walked leisurely. The high-energy ones left with many questions on how things were going to work.

As an older team member, Manny looked wowed but was very skeptical of what he heard. "The new leader always begins with words that ignite momentum, but when reality sets in, those words quickly fade away," Manny mumbled to himself. "I have been here for ten years now, and I see them come and go."

At that, Manny saw Tarasi. She reminded him of his daughter, so he always had a soft spot for her. Tarasi slowed her pace to fall in line with him.

"What did you think of the presentation?" Manny asked.

"I think it sounds great!" Tarasi replied, her voice showing her enthusiasm. "What did you think?"

"You know how these things go, promises, promises," Manny muttered. Then, he looked at Tarasi with his neck pulled into his body and shrugged. "I don't know, though. She might be different. I felt heard for the first time in a long time today when Ashanti posed your question to the group, and they chanted a response."

"That was fire!" replied Tarasi.

The agile coaches and trainers were hired as consultants to help the teams through the agile adoption based on their selection of Kanban, Scrum, or SAFe. Ashanti requested that the consultants lead discovery workshops to gather information from her teams about the pain points that were affecting progress and success, and the gain points that described what people would like to get out of the new directions.

Each pain point description would be assigned a value of 1 to 10 to describe how painful the experience had been. The scale started at 1 (the pain discomfort is manageable) and went to 10 (it hurts so bad I want to faint). A similar scale applied for each gain point description. The scale started at 1 (a small improvement would be great) and went to 10 (tear it down and build something awesome). Basil, a coach and the lead consultant, demonstrated in Figure 1 the pain and gain points.

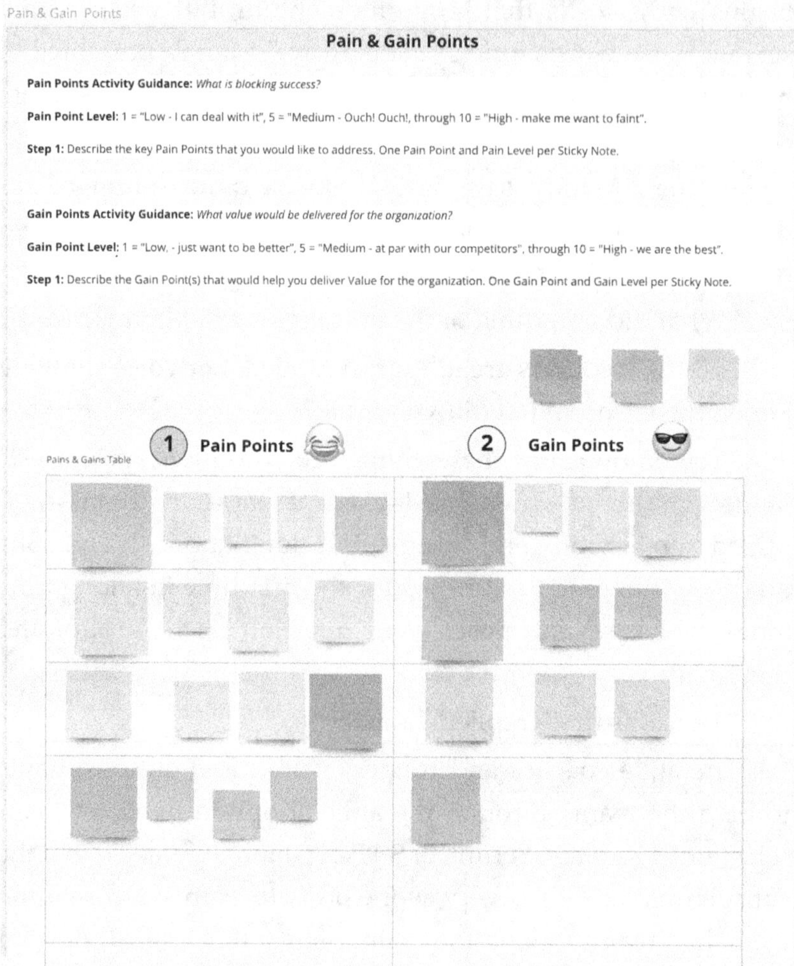

Figure 1: *Pain and Gain Points*

"How would you like the information presented?" Basil asked as he and Ashanti met together in her new office. The sweet scents of vanilla and honey had been utilized as a way to make every conversation a bit more relaxed.

"I would like the pain and gain points grouped and prioritized by the highest assigned value," Ashanti replied.

"We can then work with the teams to select the top three items to create a backlog of items to work on in the near term."

"There is a practice called Objectives, Measures, and Value (OMV) to help leaders define what is important to produce a measurable outcome," Basil explained. "OMV is part of the Agile Fluency Model used to help organizations make decisions about investments for improvements.

"Objectives are what team capabilities you would like that you don't have. Measures define how you will know the teams have those capabilities. Value describes how these capabilities will benefit you and the organization." Basil provided an illustration (Figure 2) to help Ashanti visualize the OMV.

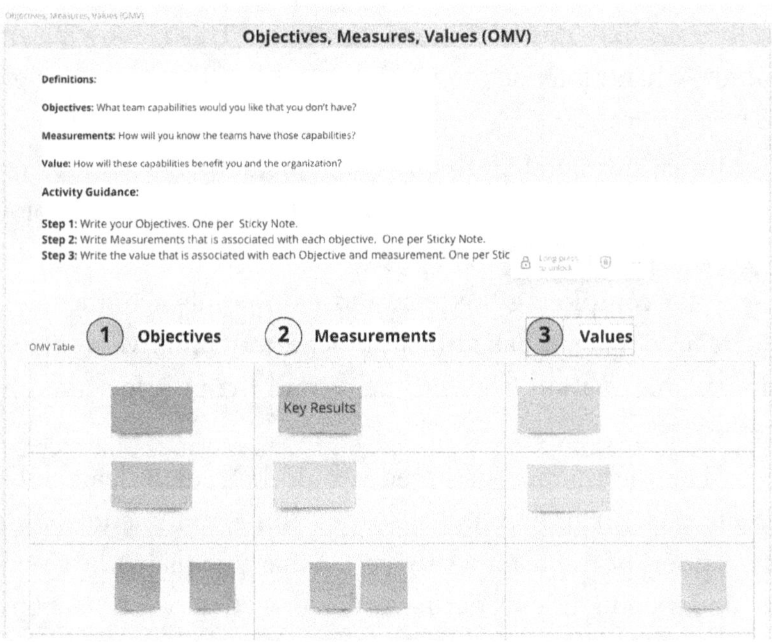

***Figure 2**: Objectives, Measures, and Values (OMV). **Adopted from**: Agile Fluency Model.*

"Perhaps we can explore how to use OMV in presenting the findings to the teams and other leaders in the organization," Basil suggested.

"I love that idea!" Ashanti replied.

The team members in Ashanti's group were enthusiastic after participating in the self-selecting team activities. Tarasi fist-bumped Manny with great exuberance, yelling, "Dude, we are on the same team!"

Since Tarasi was a user experience (UX) designer and Manny was a quality developer (usually referred to as QA, short for quality assurance), they each had a role to play in the group that was a bit different but equally important.

"You know, Tarasi, this is the second time that I felt heard in the last sixty days," Manny admitted. "This is blowing my mind. Ashanti may be the real deal, but I still reserve the right to be skeptical."

Tarasi shrugged and said, "Dude, it is in your nature. You test stuff to make sure they work. I think Ashanti will blow through all of your tests, though."

After completing the pain and gain points activities with the teams, Ashanti asked Basil to lead the initiative to gather similar information from her peers and the people served by her teams.

"The information gathered should help us to learn more about our customers and also get some sentiments of the perception of the teams' service," Ashanti explained. "I need to understand if our partners perceive that value is being delivered."

Ashanti scheduled a team meetup to share the results of the pain and gain points activities. Manny and Tarasi volunteered to share the experiences provided by the team members and helped to facilitate creating themes on the digital board. They used a digital collaboration board to allow distributed attendees to participate. The process involved inviting attendees to group the digi-notes into common themes.

One theme identified was that most work was treated as a crisis. That topic sparked a long and emotional conversation. There was pent-up energy that needed to be released, and so it was released. One of the detailed digi-notes in the crisis theme stated, "Leaders need to see people and not resources."

"Who wrote the statement on this digi-note, and can you tell us more?" Manny asked.

The team was silent for a moment as the splashes of blue through the meeting room reminded everyone to stay calm. Feet shuffled. Glances were passed around. Though Ashanti made them feel comfortable, there was still some hesitation.

Finally, Jenna stood up and said in a tense voice, "We are not just replaceable pieces in a game. We are human beings and should be treated as such. We do not have concerns about being held accountable, but we need time to evaluate the problem in order to provide a solution.

"The problems presented have a predefined delivery date without conversations with the people building the solution. My teammates and I are always on a death march, and when we push back, there is an unfair labeling of not being a team player. This is the worst experience I have ever had as a professional, and I am not sure how much longer I am able to stand this level of insanity."

When Jenna wasn't reprimanded for this honesty, there was a noticeable shift around the room as fears and hesitation dissolved in this supportive environment. The grouping of digi-notes continued, and several team members got up and bravely shared the unhealthy lived experiences affecting their ability to be happy, contributing people. The team members also shared positive experiences that lightened the darkness of negative experiences shared by the group. A positive experience shared was that people received a salary that was competitive with the market.

Basil said, "If you have a gripe, there should also be an associated hype."

The top theme selected for improvement by Ashanti's team was "be people-positive." Tarasi read the digi-note content, which said, "Focus on the contributions and potential of the people. We are more than a cost to the organization and provide value through our innovation."

As the meeting concluded, Ashanti stood and thanked everyone for being trusting and vulnerable in an open space. "I realize that this was not easy for you," Ashanti acknowledged. "Based on the challenging experiences that you have endured and shared, I hope you will lean into the changes that we will co-create as a team. We can't expect perfection. We will learn as we travel along on this journey. The destination will be defined based on our alignment as an organization and team.

"I have two very important objectives for us to pursue together, and they are to attract and retain resilient people and to practice agility. I believe these two objectives will help us co-create the be-people-positive mindset and experience.

"Please feel free to reach out to me if you have questions. The coaches are here to support your effort of practicing agility. Before I adjourn this gathering, keep in mind how you can help co-create our work environment to enable happy contributing people."

Manny linked up with Tarasi after the meeting and asked her to join him in a conversation with Basil. They approached Basil with excitement and jumped right in by asking him how they could help.

"I appreciate your enthusiasm," Basil replied. "The team of coaches will send out an invite to help, and you will be on the list."

"Thank you," they said before walking away.

Tarasi turned to Manny and asked, "Are you excited?"

He replied, "I am super pumped about what is possible."

Tarasi joined him in unison to parrot his favorite phrase about reserving the right to be skeptical.

HAPPY CONTRIBUTING PEOPLE - INFLUENTIAL LEADERS' VOICES

Diana Larsen, author of *Agile Retrospective* and co-founder of Agile Fluency Model

The Agile Fluency model has four zones, 1) focusing on value (shift team culture), 2) delivering value (shift team skills), 3) optimizing value (shift organizational structure), and 4) optimizing for systems (shift organizational culture). The approach that you're going for is either the focusing zone or you're going for something that includes those skills of the focusing zone. It's embedded in the whole model, and those are the skills that allow for becoming effective as a team. When people have that sense of effectiveness of building good products, it's very tied to the project Aristotle ideas of psychological safety and dependability and having structure and clarity around what we're working on and those five keys that they talk about.

When we have the project Aristotle experiences for our teams, they tend to have much higher job satisfaction. I tend not to think so much about happiness because that has certain connotations for certain people, but I believe that people deserve job satisfaction. They need to be pleased with their work. They need to have that sense of accomplishment. They need to have that sense of control over their work lives as a team and the autonomy that they must think about. That is true at the individual level, but it's also true of the team level.

Teams that can make decisions together and that feel they have the mastery and the skills that they need to move forward together: that's what fluency is about. We've got the right skills that match the work that we need to do, or our company is willing to invest in us gaining those skills. When you've got all that together, you have high levels of job satisfaction. I really think that the agile fluency model is a lot about creating teams that have good meaningful working experiences together and the sense that they can work well enough together that they have a positive impact in the world and that becomes important.

The agile fluency model doesn't talk about software ethics, but I think that's another conversation that's starting to rise a lot in our industry. The questions people may ponder include a) do I feel good about my work? b) does it feel like ethical work to do? and/or c) am I really serving everyone? I think the agile fluency model speaks to that happiness job satisfaction question indirectly, but it's embedded there, and it is an outcome. When people have invested in their teams to create work that feels like success and achievement, people tend to feel better about the work that they do and they're working together.

Howard Sublett, former CEO, Scrum Alliance

Happy contributing people? I've already described my ideal environment, so I'll talk about my ideal culture: One where people are free to try new things and it's okay if they fail. Not just okay, but people cheer when it happens because they tried something risky and failed early. The byproduct is that people are happy to contribute to things because they aren't afraid to speak up. They have permission to say the wrong thing sometimes and don't have to guard every word or worry that they might say something dumb. In that kind of culture, no one feels stuck; there's always a path forward.

I want the whiteboards, too, but the joyful, prosperous workplaces—the ability to work at a sustainable pace—those for me are more important than the physical space. For example, at the last two consultancies that I was with, we had an open and engaging environment. People couldn't wait to come to our Office Days, what we called our Gatherings. They were excited to learn and create together.

And we had town halls where 97% of employees would show up every single month. It wasn't the administration telling them something; it was the team coming together to share their stories and learn from one another. Everything—even small things like our employee handbook or expense policies—was open for discussion, interpretation, and revision.

I would count both of those companies as great experiences for learning how to create those cultures and live within them.

Marty Nelson, Alchemy Code Lab founder:

Meaningful work, shared purpose, creativity. In my own life, I struggled in college because there was no one being served by the work I was being asked to do. But I remember one of my early jobs as a developer working at a company that did legal services, and we were building a system to process claims from the 9/11 World Trade Building–looking through the data and the responses were so emotionally strong that the loss and tragedy was palpable. I saw this connection from the work we were doing to significant things going on in the real world and real people being impacted. People need to feel that their work has meaning and produces impact.

Having a sense of purpose within the team or organization is also important. It wasn't until I had my first job as a development manager that I was aware of how dependent success was on the work of the team. It was out of my hands. I had to rely on and trust everyone to do their jobs. But what I noticed was that the successes felt better; we got more out of each win. That team delivered in six months a project that had floundered for almost two years. We all did our part, and everyone felt really good about that.

Dave West, CEO, Scrum.org:

The most important thing is "purpose." We need to know why we do something, what our ultimate purpose is. It is important that the purpose is not abstract, but very tangible and real. For example, I was working with a big pharma sales and marketing IT group—one product they provided was a social network for people that suffered from multiple sclerosis (MS). Their purpose at an abstract level was to connect individuals who suffer from MS with caregivers, and for caregivers to find patients to help them deliver better care. But building a Facebook for MS patients was not a tangible mission. They took pictures of real MS customers and built personas around them. They made their needs very tangible and described how the work the team was doing made their lives better. Every Sprint Review asked, "How does this help Sally [a persona name]?" Everyone was motivated, focused on the outcome. It was not just a set of features on a Facebook knockoff; it was a way for Sally (and others) to get advice on her pain at 3 a.m. It was a feature to allow Simon to get better feedback on a protocol to help Sally, etc.

And yes, to quote Dan Pink after *purpose,* "autonomy (or being trusted) and mastery (getting better at something) are key." We build environments that enable that—that then creates happiness—which in turn improve outcomes. To quote Richard Branson, "Happy people make happy customers, which make happy shareholders" or to badly change a quote from *Lion King,* "It is the cycle of value."

BELONGING LEADS TO HAPPINESS

Another way that people experience happiness is through a sense of belonging. A variety of studies have shown that belonging is essential to happiness and satisfaction with life. People are born into groups and spend many years being members of varying dynamic collectives that include work, sports, and hobby groups. Social integrations promote well-being by giving individuals a sense of meaning, purpose, and support during stressful times in life. Not every social integration generates happiness.

Creating a generative belonging workspace means that people are resilient in their ability to give birth to new ideals of who they are. This means individuals have a sense of purpose. People are allowed to bring their whole self to work and experiment to improve who they aspire to be. The space of belonging is for everyone in the organization to participate and co-create a generative work environment. We spend approximately 50% of our awake hours in the workplace, so feeling connectedness to a group or community by being accepted by your people increases a sense of happiness. Happier people are 12% more productive at work (Oswald, Proto, and Sgroi, 2015), and we may be able to reason that it is the same percentage in life.

Happiness is a catalyst to improve individual and team productivity and innovation. Given this knowledge, perhaps more organizational leaders would strive to create environments that are conducive to more happy contributing people. Why not focus on helping to increase the bottom line of increased profitability? Happy contributing people help your organization deliver value.

BE PEOPLE-POSITIVE

People celebrate high-performing teams, and they are often given the best assignments and admired for the outcomes achieved. High-performing teams are known for being self-organizing, collaborative, and domain masters. Emergent leaders are the cornerstone of high-performing teams, which are the team dynamics desired by any leader in the organization.

People are the lifeblood of the organization. This is a statement that I have heard and read about from many C-level executives during my corporate experience. I question the sincerity of this statement based on what I have witnessed and experienced of how people are treated in businesses. People are treated and referred to as resources. We should not refer to people as resources. When people are referred to as resources, they become objects and lose their humanity. The definition of resource is "a stock or supply of money, materials, staff, and other assets that can be drawn on by a person or organization in order to function effectively" (*Merriam-Webster Dictionary online*). Even in this definition, the word "staff" is included, which should be translated to mean people skill. The education and training I have received placed the descriptor "resource" in my vocabulary, and I have been guilty of objectifying people as well.

According to Dignan (2019), an organization's leaders must adopt a people-positive mindset. I am borrowing the "People Positive" title from Dignan because I found it to be relatable. People are capable and self-organizing, worthy of trust and respect. McGregor (2006) expressed in the *Theory Y* research and discovery that the full capability of human beings is being able to self-direct and that they are motivated by purpose and mastery. The *Theory X* management style holds the view that people are not trustworthy and need to be directed (McGregor, 2006). Daniel Pink's view is people are motivated when they have autonomy, purpose, and mastery. Maslow and Pink share a common mindset of a people-positive context in that people are capable of providing value without the constant oversight of management. Dignan (2019) stated that if you treat people as all-stars, then they will become all-stars. Expect and assume the best of people to obtain the people-positive mindset. Positive human value produces positive business outcomes (Schmidt, Rosenburg, and Eagle, 2019). Help people to go beyond their imposed constraints to achieve a personal level of awesomeness.

An important aspect of building team culture is developing a learning community. Personally, I engage in daily learning through podcasts, audiobooks, and conversations with people. Being people-positive is about creating a space for people to learn and grow. Each month during my team meeting, we engaged in team building. Learning together is an important team activity to share in a common topic that develops focus on a shared language, understanding, and purpose. I wanted the team members to create purpose as a team and individually. One of the first books we read together was *Start with Why* by Simon Sinek. The feedback from the architect on the team was that he rarely reads non-technical books and it was enlightening to discover the importance of discovering "Why." The twenty-six-member team began developing a closer relationship because they now had a shared sense of purpose.

In an experience void of happiness early in my career, I worked for the now defunct Arthur Andersen (AA) and Andersen Consulting (now Accenture). The AA world headquarters was located at 69 W. Washington Street in Chicago, IL. I could look out the office windows and see the Daley Center and Plaza across the street located at 50 W. Washington Street. I have vivid memories of watching the St. Patrick's Day parade from the top floor of the building, moving on the cross streets of Dearborn and Washington in the spring in Chicago as I took a break from working. I was assigned to a project that would consume my every moment for almost a year. It was truly the infamous dreaded death march. The delivery date was so aggressive that we were placed on a mandatory seven-day work schedule for more than nine months. The schedule was twelve-hour days Monday to Friday, ten hours on Saturday, and eight hours on Sunday. We began in the late spring, and I could not remember what I did that summer. I called it the "lost summer." I remember a coworker having a mental break as she was screaming and standing on a desk in her office one late night. The partners called a meeting the next day and told us to take the next weekend off but be ready to get back on the same schedule. We would eventually finish the software project, but we were all toast. After that experience, I began looking for another job and left as soon as possible. I can tell you that experience did not make me (or any of the people on the team) a happy contributing person.

Happy contributing people are autonomous, have a purpose, develop pertinent skills and capabilities, and are encouraged to bring their "whole person" to work. My great-grandmother, Ma Ruby, would often say to me, "No two horses run alike," in her Caribbean colloquial dialect. She meant that each person has unique characteristics and awareness of this reality was important. The idea of happy contributing people is that people are happy based on positive circumstances and do the best possible work based on the contribution of individual capabilities to create a high-performing team. A team of people functioning as happy contributors creates positive value for the organization. This is not to say that the working environment is one of cotton candy and ice cream. People hold each other accountable with expectations to deliver value that benefits each other and the organization.

EMPATHY AND COMPASSION

In the modern workplace, it is now acceptable to talk about feelings so we can demonstrate empathy and compassion for our coworkers. Recognizing the humanity and needs that we share daily can produce high-performing teams of people that make coming to work enjoyable and, in most cases, produce happy contributing people. We can create an environment that influences the leaders and others to behave differently without the need to dehumanize each other. We can still compete and work toward personal goals while sharing empathy and compassion with our team members.

Recently, I worked with another agile coach at a multi-billion dollar company. We will call him Tommy. Tommy was hired two months before me and assumed the role of being the "program coach" with aspirations of being the "portfolio level" coach. A decision was made by the company's leadership six months after I started to reduce headcount and have only one coach. Fortunately, I was chosen to be the coach because of the intangibles that I brought to the table of training, coaching (business and executive, agile enterprise, etc.), and leading workshops. Personally, I felt compassion for Tommy, my fellow coach, and reached out to express my empathy for the decision affecting his employment. I could connect with his anguish of being released after helping to launch the agile adoption for this company.

The first thing I did was to call him by phone to see how he was doing and offer my support with contacts that I have in the area and act as a professional reference. We did not always agree on an approach to provide agile training and coaching.

We had passionate exchanges of words and held firm to our convictions. But Tommy is a fellow human being and deserves empathy and compassion. Based on my own work experiences, I have been there before. I knew how it felt to be replaced after helping an organization to launch a new way to work that would improve the ability to be more responsive to market changes and deliver customer value faster to the market. This is not a good feeling, but I certainly learned how to be more empathetic toward other people after that experience.

ACCEPTING THE WHOLE PERSON

Most corporations only want the work persona (self) to show up at the office. They do not want the human self that represents the whole person. You often hear managers and coworkers say to leave the emotional "you" at home. Can you imagine the suppressed innovations that never see the light of day because we are constrained to only bring some parts of who we are as humans to work?

As a former manager of people in a corporate job, my goal was always to get to know people beyond the work persona that showed up at the office Monday through Friday. My curiosity led me to learn about their families, hobbies, and aspirations. During the first one-on-one meeting with a new person on the team, I would ask powerful and open-ended questions that provided insights into what that person valued and how they would like to be recognized. The questions about what people valued included family, money, new knowledge, and more. Additionally, it was important to learn about how they would like to be recognized. Some people are embarrassed by public recognition, so I wanted to be mindful of the whole person. Letting people know that we want a culture that celebrates excellence was also important for team culture.

The first question that I would ask the people on the team during one-on-one meetings was how they and their families were doing personally. Knowing the stresses and joys of people's lives helped me to sense and respond accordingly during the time we spent together. We often would go for walks as part of the one-on-one meeting to remove the barrier of the desk or table that would symbolize hierarchy. The goal was to build deep relationships with my team members so we could be transparent and have hard conversations without them becoming antagonistic. As an aspiring servant leader, my role was to help the people entrusted to my care to achieve awesomeness.

HAPPY CONTRIBUTING PEOPLE ANTI-PATTERN

Anti-patterns are poor solutions to recurring design problems. Anti-patterns are harmful behaviors and practices that are intended to improve a process. Anti-patterns are behaviors like micro-managing a person's work and not allowing autonomy. Empathy for people is never considered or is completely ignored during the design and application of the anti-patterns. The focus is on the fact that there is a problem and never "why" the problem exists. Discovering the "why" will help design experiments that can reduce the anti-patterns and provide generative options that create an environment for happy contributing people. The people responsible for reducing the anti-patterns are leaders as well as team members. The happy contributing people anti-patterns limit the ability for teams to grow into high-performing teams.

Evaluating People's Performance

Performance measurement is a fact of life for full-time associates, consultants, and partners working to support a corporation. Achieving awesomeness is enabled by mindfulness, lean thinking, and agile practices. The practice of mindfulness is awareness of your present needs and those of others. The story of the golden apples describes how we are distracted from the present moment by many events that take away the ability to achieve awesomeness. The story about the golden apples comes from Greek mythology. As the story goes, in one race, Hippomenes (or Melanion) was given three of the golden apples of the Hesperides by the goddess Aphrodite. When he dropped them, Atalanta stopped to pick them up and, in doing so, lost the race. Atalanta had never lost a race before because of her focus on the moment. The beautiful golden apples distracted her. The central theme of the story is to not lose focus on what is important, which for us is the people contributing to satisfied customers and the thriving business.

If You Mean People, Say People, Not Resources: Stack Ranking Anti-Pattern

People are the lifeblood of the organization. When people are not treated with respect, equipped with the right tools, and afforded the environment to thrive, it is non-productive. Providing team members with what is needed to do the job well is the responsibility of the organization's leaders. Recently, while working for a bio-technology company, the director of research and development and a consultant decided to use the controversial forced ranking model to determine how people stacked up in the IT department. General Motors practiced forced ranking from 1980 to 1990. The Ford Motor Company used forced ranking until 2001 and abandoned the practice after a lawsuit settlement for $10 million. Yahoo, the internet company, had its own version called "stack ranking" and has since abandoned the practice. Motorola used the management performance appraisal method from 2001 to 2003. Many companies used other language to mask the forced ranking management practice to minimize the lawsuits that many filed for gender, age, and race discrimination.

In the scenarios described above, people were treated as resources that could be discarded. The mindset was to create harmful competition that removed individuals evaluated at the bottom of the rankings. Adopting a forced/stack ranking pattern is a function of poor leadership skills from someone who lacks the imagination to create a space for people to thrive by contributing to the organization's success. As you can imagine, the remaining people were distrustful of the leadership and faked having authentic relationships with the people in leadership as well as with each other.

Measuring People's Capability Through the Reference Story Point Measurement Anti-Pattern

In a recent agile transformation journey experience, a senior vice president (SVP) that I worked with created a reference-story model to evaluate software engineers based on a minimum of five years working experience with a specific technology stack to determine how many Story Points (SP) each software engineer should deliver in a sprint. A sprint duration is commonly two weeks. A Story Point is a relative estimate that uses the Fibonacci sequence (1, 2, 3, 5, 8, 13, …) to identify if one unit of work is smaller or bigger than another. A team of developers with similar education and experience duration begin by identifying the smallest unit of work to assign a value of 1. The team then collaborates to identify the largest unit of work and assign a value of 13. The middle value is somewhere between 1 and 13, and we can assume that the value is 5.

The goal was to establish an objective expectation, based on experience and education, that everyone should relatively perform the same. My great-grandmother, Ma Ruby, would often say to me, "No two horses run alike." This SVP's expectation was misguided and created an environment of fear. The trust factor was eroded, and people were not transparent about the work's progress. People began to hide failures and found ways to blame others. The blame game was in full motion. A key tenet of agility is to have respect for people and culture. The SVP clearly did not align with this tenet, and his experiment failed miserably. Evaluating people's performance using a stack ranking model was applied in this environment as well. The result was that almost every three months, there was a great purge and the aroma of fear was present in the work environment.

The SVP became aware of the impact of his intention through his direct reports and peers and had the courage to walk back the decree. With coaching and time, the SVP changed his perspective and began to realize that his solution to a skills gap issue was not to create a false narrative that implies that if we have the same education and length of time in a specific role, our results should be the same. Being open to acknowledge when experiments are not working is an important characteristic of assigned or emergent leaders. Demonstrating the courage to change and perhaps do a public failure bow goes a long way in illustrating vulnerability and humility. The public failure bow is demonstrated when a person stands up, bows, and says, "I failed, and I hope to learn and grow from this experience."

THE VALUE OF BELONGING, DIVERSITY, EQUITY, AND INCLUSION (BDEI)

The diversity, equity, and inclusion (DEI) conversation is essential in a multi-identity society like the United States of America and other countries in the world. As our society becomes more and more diverse, what steps are organizations taking to ensure there is a place for all who are capable of contributing to disruptive innovations? People need to be loved—simply meaning they need kindness and patience as they work to be value contributors in the organization. Identity such as age, gender, race or ethnicity, religion or spirituality, socio-economic status, accessibility, and sexual orientation should not limit a person from dreaming to achieve their level of awesomeness.

We live in a world that is diverse and complex with a need to include all brains and hearts that can contribute to the innovation of new products and services. The absence of DEI creates friction in our businesses and society that are not productive. Schell (2017) said, "Diversity is a mindset that needs to touch every aspect of a company so that organizations ensure everyone embraces and contributes to the DEI culture."

The Value of Belonging

We experience belonging when there is a sense of acceptance, inclusion, and identity within a specific group. Since birth, most humans have been guided to belong to our family and community. Belonging should be central to help people envision identity. Belonging is a basic human need. It's an important part of who we are and how we find identity.

Instinctually, our need to belong draws us into relationships that are helpful for spiritual, emotional, and physical development. The experience of healthy belonging in a positive relationship and environment helps us to feel valued. Belonging can be a generative experience that is beneficial for everyone. Some relationships don't always provide the beneficial experiences that we desire and can be deadly.

Everyone needs to belong to a community or organization. Belonging belongs to everyone in a work environment and community. One group of people does not hold a belonging monopoly over another group of people. The participation of all members of an organization is required to create a healthy space of working. Belonging helps to increase acceptance, inclusion, and compassion.

The Value of DEI

DEI has been center stage to create an inclusive work environment and society. Diversity and inclusion initiatives are often owned by the CEO or managing director. Oftentimes, the ownership may include the board of directors to satisfy an organization's strategic initiative. The delivery of DEI can be assigned to internal teams and external experts to form a collaborative partnership and team to make all people a priority. The sharing of the DEI ownership with members of the organization permits open discussions that can lead to better understanding between the people affected by the absence of a DEI culture.

In 2016, *The New York Times* article "What Google Learned from Its Quest to Build the Perfect Team" highlighted the significance of Google's application of Amy Edmondson's discovery that people working in a psychologically safe environment produce high-functioning teams. The psychological safety concept was introduced in Edmondson's 1999 research paper at Harvard University, titled "Psychological Safety and Learning Behaviors in Work Teams." In 2015, Google established that psychological safety is the foundation for their highly effective teams. The teams had a culture in which people felt comfortable taking a risk without feeling insecure or embarrassed.

Psychological safety is a term used frequently in the agile community. Psychological safety is feeling comfortable sharing ideas without fear of being shamed or belittled by your peers. I would say that DEI is a psychological safety concern that needs to be championed by agile coaches and leaders. However, we need everyone in the organization to become aligned with this way of thinking and behavior to achieve psychological safety and DEI agility.

The need for allies from the majority race or class to enable DEI effectiveness cannot be overstated. People willing to step into the role of an ally to improve humanity create a harmony that enables the organization to deliver value to customers and markets. What can an ally do to make DEI a reality? It begins with empathy and compassion, while walking shoulder to shoulder.

If the organization leadership's strategic goal is to create a world class business that satisfies the customers' needs and attracts the best minds, then DEI will be a priority. People want to work for progressive companies that practice fair treatment, access, opportunity, and advancement for all. The absence of DEI in an organization is waste (known in lean thinking as muda) and robs the people of cultural riches and innovative ideas.

Belonging, Diversity, Equity, and Inclusion Intent and Outcomes

A diversity and inclusion initiative should be treated as an outcome with clearly defined and measurable goals. The organization's leaders are accountable for setting the outcomes for BDEI that are aligned with strategic goals. The outcomes should have a vision to guide the value delivery team(s) assigned to lead the BDEI change for the organization. The BDEI value delivery team should incorporate an agile practice to enable iterative deliverables, which allows the team to pivot as needed to ensure quality. The organization's intent is to treat BDEI as a business value and intellectual property. The following are seven levers to assess the success of the BDEI outcome:

- Engaged leaders (C-level and managers) are supporting allies and sponsors
- A vision tied to a strategic outcome with measurable goals
- A value delivery team comprising internal people and external experts
- A budget to pay for the BDEI initiative
- Lean thinking and agility tenets used as a value delivery framework
- Changed organization policies and practices
- Shifted culture with a focus on BDEI

The BDEI assessment levers will help the BDEI value delivery team and leaders when reviewing the outcomes of the initiative. The assessment should be made visible to the organization for people to begin the alignment process. The victories, as well as the challenges, are equally important to share.

Diversity and Inclusion Pillars

DEI is a continual journey that requires the positive intent of a people-centric organization. Leaders must be aware, people-centric, socially responsible, and allies, and able to facilitate the empowerment of marginalized people to ensure all brains and hearts are on deck to continually improve business innovations and the society we share. The key actors in BDEI are the partner, agent, and target.

The Office of Multicultural Affairs at USC defined a *partner* (formerly referred to as *ally*) as "a person of one social identity group who stands up in support of members of another group; typically, a member of a dominant group standing beside a member(s) of targeted group; e.g., a male arguing for equal pay for women" Goldbach (2017).

Goldbach (2017) defined agents and targets as follows:

> Agents are members of dominant social groups privileged by birth or acquisition who knowingly or unknowingly exploit and reap unfair advantage over members of the target groups.
>
> Targets are members of social identity groups who are discriminated against, marginalized, disenfranchised, oppressed, exploited by an oppressor and oppressor's system of institutions without identity apart from the target group, and compartmentalized in defined roles.

The practice to guide and support the partner, agent, and target engaged in the BDEI agile journey will include four pillars: 1) awareness, 2) building, 3) elevating, and 4) sustaining to help organizations develop a sustainable practice of employing human beings regardless of their identity.

Figure 3 illustrates the BDEI enablement pillars that provide a framework for the organization to engage in creating an environment of acceptance and innovation.

Identity, privilege, and power are the key factors that limit the ability for BDEI to become a natural stance in organizations. The identities include age, gender, race/ethnicity, religion/spirituality, socio-economic status, accessibility, and sexual orientation. Privilege is the rights, benefits, and advantages enjoyed by a person or group over another person or group. Privilege also refers to groups or people with historical advantages that include economic, freedom, and power. Power is the privilege granted to some people to make decisions about the well-being of a marginalized group of people. Power is often inherited based on family tradition, socio-economic status, and gender.

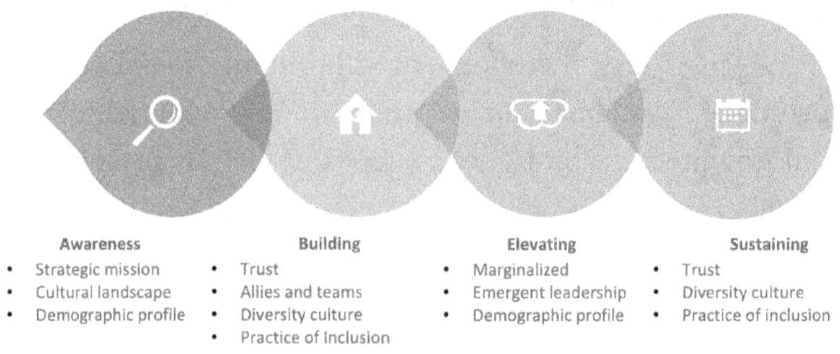

***Figure 3**: BDEI Pillars*

The goal is to arrive at a high cultural competence that reflects the words of Dr. Martin Luther King Jr., which is to evaluate a person based on the content of their character. The four pillars of the practice are intended to equip the BDEI value delivery team to actively tackle the challenges of BDEI in the organization. The Kanban board would be the appropriate lean tool to make work visible while implementing the four pillars.

The simplicity and power of the Kanban board will allow the BDEI value delivery team to present work as 1) backlog (list of work), 2) to-do (prioritized work), 3) doing, and 4) done.

Awareness

The awareness pillar consists of three areas of focus that include strategic mission, cultural landscape, and demographic profile. The strategic mission is associated with the organization goal that is owned by one or more members of the C-level group or board. The cultural landscape is the state of BDEI in the organization, which includes beliefs, attitudes, and norms. Demographic profile is the composite of the people's identities throughout the organization. Awareness begins with mindfulness by members of the organization to be on the lookout for instances of psychologically unsafe situations for people in a target group as well as the agents. As agile coaches, we are responsible for making work humane; we owe it to teams and organizations to create psychological safety for team members with diverse identities.

Building

The effort of building and developing a people-centric organization is not easy because often we are only focused on the bottom line and miss the opportunity for emergent leaders from a diverse group to leverage a growth opportunity. Fear, which is often the internal engine of oppression, limits the ability for members of the organization to change and enable a diversity culture. The building pillar emphasizes developing trust, partners and teams, diversity culture, and the practice of inclusion.

Trust is essential for the leaders, BDEI value delivery team, and recipients of the proposed BDEI culture to experience together. Trust includes three attributes:

1. Empathy – Ability to feel and connect with another person's experience
2. Integrity – Authentic and sincere stance
3. Ability – Capable of delivering what was promised

Trust will work best when there is an understanding that:

- Trust works both ways
- There is authenticity and people are met where they are

The partners and BDEI value delivery team(s) are responsible for partnering with external experts and members of the organization to create a diversity culture and to create a practice of inclusion. The allies can be members of the DEI value delivery team and can also be other members in the organization. The practice of inclusion is an activity that may require a partnership with human resources (HR) to ensure compliance with regulations and corporate policies. The HR group can be an important ally to schedule and track training for the members of the organization.

The BDEI value delivery team will include goals and methods to validate quality as part of the planning activities. The objectives and key results (OKR) method is one tool to validate the achievement of planned outcomes. OKR is a method for defining and tracking objectives and outcomes. The OKR method was developed by the Intel Corporation and later adopted by Google.

Elevating

The BDEI outcome must be more than a development of practices and a set of feel-good mantras. A positive environment is needed to elevate the marginalized that is measurable and visible. First, begin with elevating awareness of the BDEI culture shift initiative throughout the organization. Why not leverage the talent of the marketing department to create visually engaging items to promote the BDEI initiative? Social media is another tool available in the organization to elevate awareness and participation. Imagine the elevated human dignity that will be achieved by promoting fairness, access, and opportunity to grow. Additional training may be required to help leaders with direct reports to remove the blinders of unconscious bias to create a success path for all regardless of identity.

Sustaining

As the organization begins to treat BDEI as a value contribution to people, the importance will include sustaining trust, a diversity culture, and the practice of inclusion. Developing learning communities to self-organize and sustain the BDEI culture may develop stronger allies. A system that recognizes individuals who are allies and advocates for the BDEI culture would energize others to participate. Gamifying the sustainability of the BDEI culture can lead to other optimizations identified by the learning community. Promoting transparency by capturing and presenting metrics for the management team to review and be accountable for will ensure focus is maintained on the BDEI initiative. Finally, reviewing the OKR metrics on a monthly basis will show if the goals identified are being met or if they need to be modified. Practice the inspect and adapt actions of agility.

HOW ARE MY PEOPLE DOING?

In the technology product development industry, especially when information technology is involved, many measurements are designed to capture what people are delivering. In companies using the agile framework, you will find burndown and velocity charts to show how fast teams are producing work items. Often, leaders push back on measuring the happiness and engagement of the people delivering customer value. The rationale is the measurements are too squishy! I could not disagree more with this way of thinking because we are losing sight of the value that happy contributing people provide to customers and the organization.

The reality is leaders are still evaluating people through the lens of Taylorism of the early 1900s through the 1980s. Frederick Taylor's scientific management theory promotes the idea that there is one right way to perform work. The translation is there is one way to measure people's performance. My guidance is a proposed "Yes and" approach that includes a limited performance measurement such as a velocity chart and includes the people's happiness and engagement experiences measurement to give visibility into the whole person delivering value for customers and the organization.

People Happiness Experiences

The people's happiness experiences measurement is one way that I propose to capture the key attributes that highlight people's happiness. Happiness is based on experiences and not just emotions. Figure 4 illustrates the nine values that contribute to people's happiness. The happiness experience survey helps individuals and teams understand their happiness at work through the lens of fun and people motivators (autonomy, mastery, and purpose), and coupled with the five Scrum values (commitment, focus, respect, courage, and openness).

***Figure 4**: Team/Individual Happiness Experiences Measurement*

People Motivators

Pink (2009) described autonomy, mastery, and purpose as key indicators that motivate people. Pink (2009) further explained autonomy, mastery, and purpose as follows:

- *Autonomy* – People need autonomy over tasks (what they do), time (when they do it), team (who they do it with), and technique (how they do it). Companies that offer autonomy, sometimes in radical doses, are outperforming their competitors.
- *Mastery* – Mastery is a mindset: It requires the capacity to see your abilities not as finite, but as infinitely improvable. Mastery is a pain: It demands effort, grit, and deliberate practice. Mastery is an asymptote: It's impossible to fully realize, which makes it simultaneously frustrating and alluring.
- *Purpose* – Within organizations, this new "purpose motive" is expressing itself in three ways: In goals that use profit to reach purpose; in words that emphasize more than self-interest; and in policies that allow people to pursue purpose on their own terms.

Scrum Core Values

Scrum is a framework developed in the 1990s and made popular when adopted by technology product development teams to change the approach to delivering customer value. The Scrum values represent behaviors and beliefs that enable teams to perform at an optimal level. The Scrum values include the following:

- *Commitment* – the state or quality of being dedicated to cause, activity, etc. (commitment to the team, quality, collaborate, learn, do the best they can, the Sprint Goal, act as a professional, self-organize, excellence, agile principles, creating working software, etc.)

- *Focus* – focus on expertise (Scrum roles), focus on what is important now (time boxing), focus on what we know now (Sprints, Sprint Goal, Daily Scrum)
- *Openness* – the empiricism of Scrum requires transparency, openness, and honesty
- *Respect* – respect for people, their experience, and their personal background. Respect diversity, respect different opinions, and respect each other's skills, expertise, and insights.
- *Courage* – it requires courage to keep to the Scrum framework and the agile principles. Do the right thing when it is hard.

Fun

Fun stands alone because it is the expressed enjoyment of the individual or team members. The cause and effect of fun is subjective and is subjected to the lived experiences of each individual and team. Reflecting on my experiences working in my community and with clients, fun shows up when there is laughter and connection to common beliefs and goals. One way to describe fun is "I found joy in my work and connecting with my team members."

The people happiness experience survey consists of nine questions. Each question uses the five point Likert scale response options that may include strongly disagree, disagree, neither disagree or agree, agree, and strongly agree. Other response options can be substituted for the Likert scale response options described here.

The people happiness experience survey includes the following questions:

1. *People Motivators* – I experienced Autonomy in my team to work at a sustainable pace.
2. *People Motivators* – I was able to increase my Mastery working in my team.
3. *People Motivators* – I understood my Purpose in the work I performed in my team.
4. *Scrum Value*: I was able to Focus on what was important to the team.
5. *Scrum Value*: I was able to practice Openness by being transparent and sharing honestly with my team.
6. *Scrum Value*: I showed Courage to do the right thing or share my opinions even when it was difficult.
7. *Scrum Value*: I showed Commitment to achieve the goals of the team.
8. *Scrum Value*: My Scrum/agile team members showed Respect for my capabilities, opinions, and insights.
9. *Fun*: I had Fun working with my team.

The people happiness experience survey responses are measured using the following value ranges:

- 80%-100% (Happy) – support the team's focus to maintain current patterns to improve
- 60%-79% (Moderately Happy) – monitor the patterns observed, influence to grow
- 59% and below (Not Happy) – seek to understand and coach toward improvement

People Engagement Experience

In the new order of things, workforce development is center stage in the conversation to upskill members of the organization. It is estimated that un-optimized training leads to financial and opportunity loss for companies, as much as $550 billion in productivity loss to organizations globally (Solis, 2015). We should not lose sight of the execution engagement that produces value for the organization.

The people engagement experience survey consists of seven questions. Each question uses the four point Likert scale response options that may include: exceeds expectations, met expectations, below expectations, and did not meet expectations. Other response options can be substituted for the four point Likert scale response options described here.

The people engagement experience qualitative survey helps team members understand the level of engagement by probing with the following inquiries:

1. Understands the customer requests
2. Takes initiative to lead without being asked
3. Recommends solutions to problems and provides innovative ideas
4. Contributes to technical and other conversations
5. Meets Say – Do (Stories planned vs Stories done) working agreement
6. Supports Dependent Team(s)
7. Builds rapport and trust

The people engagement experience survey applies the 360-feedback model that helps team members use a blind selection practice to randomly provide feedback to each other. Team members receive confidential and anonymous feedback from peers and others on cross-functional teams who may also contribute insights. The people engagement experience survey responses are measured using the following value ranges:

- 80%-100% (Engaged) – support the team's focus to excel
- 60%-79% (Moderately Engaged) – monitor the patterns observed, influence to grow
- 59% and below (Not Engaged) – seek to understand and coach toward improvement

This model helps to build trust between the team members. The frequency of this event usually takes place every two or three months. High-performing people contribute to high-performing teams, and the work of leaders will be to create transparent methods that influence aspiration and engagement in this global pandemic. Figure 5 illustrates the visualized results of a people engagement experience survey.

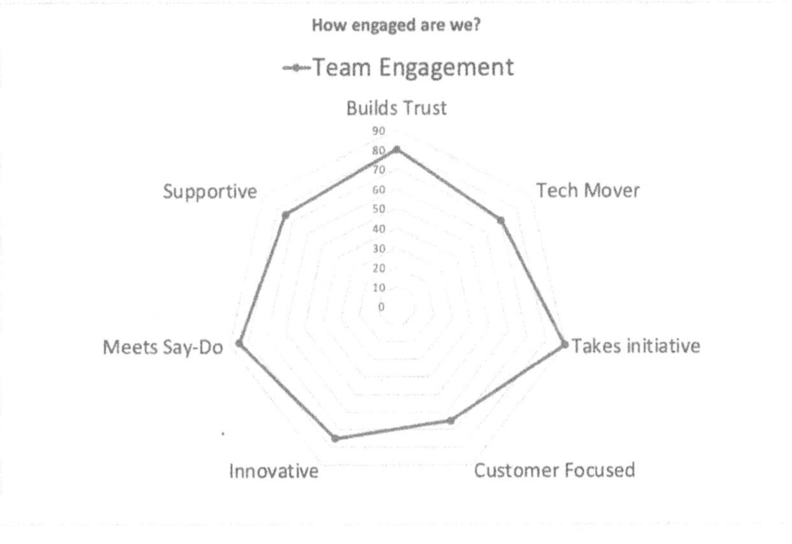

***Figure** 5: Team/Individual Engagement Experiences Measurement*

SUMMARY

Management is about human beings. Its task is to make people capable of joint performance, to make their strengths effective and their weaknesses irrelevant. – Peter Drucker, management thought leader and author

Focus on providing an environment that enables the people inside and outside of the organization to thrive so they may be happy contributing people. The people outside of the organization include partners and customers. They are vitally important for the success of your organization and its ability to thrive in a global VUCA business environment. When they are happy as well, good things will follow.

Be aware of the happy contributing people anti-patterns that will adversely affect the people responsible for satisfied customers and a sustainable, thriving business. The anti-patterns are knee-jerk reactions to a problem that can be mitigated with solid design patterns. Start with empathy and seek to understand what is causing people to not perform at their best. Ask what you can do to help them achieve their level of awesomeness.

The belonging, diversity, equity, and inclusion practice is essential in a growing global and diverse workforce. My guidance is to start with belonging so that everyone in the organization can be included in co-creating an environment for people to belong and be happy contributing people. Don't short change the DEI practice, but focus on ensuring that the underestimated people can also contribute to the positive desired outcomes. All hearts, minds, and hands are needed to succeed in this global VUCA world.

The COVID-19 global pandemic has moved the trust walk needle to enable knowledgeable workers to do their jobs remotely. In years past, this was a limited offer with much skepticism by leaders. The great resignation changed the context of what leaders are willing to do to ensure enough capable people are available to support strategic goals. The great resignation is the result of actions by many workers who left their jobs in search of better ways of working. This resulted in a shortage of skilled workers to fill roles that would allow organizations to execute strategic goals.

Assigned and emergent leaders must shift their leadership styles to be generative agile leaders. The GAL infuses great values that allow humans to realize these thoughts of Peter Drucker's: "*Make [people's] strengths effective and their weaknesses irrelevant.*" The GAL values provide a set of beliefs to guide the people directly entrusted to our care and amplify our influence with others. The GAL values include the following:

1. High cooperation – Focus on we
2. Messengers enabled – We trust you to achieve our goals
3. Risks are shared – Win and lose as a team
4. Bridging encouraged – Partner and have conversations to learn more
5. Failures lead to inquiry – Ask, "What did we learn?"
6. Novelty implemented – Run the experiment and evaluate the hypothesis
7. Abundance mindset – We have capacity to share with others

Fundamentally, leaders want to hire and retain resilient people and practice agility. Resilient people proactively respond to opportunities and threats by discovering a path forward during good and difficult times. The desire to practice agility adds a superpower to resilient people's capabilities to self-organize and innovate, which supports satisfied customers and a thriving business.

People are worth it, and so are you.

Satisfied Customers

Customers have the freedom to hire and fire our products and services.
— Clayton Christensen

INTRODUCTION

The ultimate sign of a satisfied customer is loyalty to a brand or product. A loyal customer will invest time to share the experience with a specific brand or product with family, friends, and the world for free. This phenomenon is called social proof. User social proof is when your current users recommend your products and services based on their experiences with your brand. An example of social proof may include praises on social media or positive ratings on review sites.

Another method of social proof is the wisdom of the crowd. This type of social proof is when a large group of people is seen to be endorsing your brand. An example is having thousands to millions of customers or followers on your social media profiles. The wisdom of your friends is another type of social proof when people see their friends approve of your product. An example is people seeing their friends use your product or follow you on social media.

Customers want what they want now and have a short memory about what value was delivered. So, you may hear them singing the chorus of the Janet Jackson song "What Have You Done For Me Lately." Clayton Christensen (2016) said, "Businesses really need to home in on the progress that the customer is trying to make in a given circumstance—what the customer hopes to accomplish. This is what we've come to call the *job to be done*." The understanding of the customer job to be done is the beginning of the journey to come close to having satisfied customers. You may ask how you can understand the customer's job. You begin with empathy, ask transformative questions, and observe behavior with customers interacting with your products or services.

LEADERS WANT INCREASED BRAND LOYALTY

My great-grandmother, Ma Ruby, would often tell me that a bird in hand is better than two in the bushes. Managing the relationship to retain satisfied customers is a better business proposition than having to acquire new customers. One immediate measurable outcome to the organization is maintaining a high customer lifetime value (CLV), which indicates the average customer's revenue generated over the entire relationship with your company. The CLV of the grocery store in your neighborhood is probably high. For example, let's imagine a grocery store customer spends $100 per visit and comes in an average of once every two weeks (26 times per year). The CLV calculation is the average order value multiplied by average number of transactions per period multiplied by the customer retention duration. The grocer CLV is calculated by multiplying three numbers—$100 (order value) x 26 (visits/transactions) x 1 year (retention duration)—for a value of $2,600 annually. The key advantages of understanding CLV are improving customer retention and avoiding customer attrition; increasing repeat business; encouraging higher value sales through incentives; and increasing profitability. Customers loyal to the grocer brand will continue to contribute to revenue and profitability year over year.

Having happy contributing people in your organization is critical to sustainable and increased customer satisfaction. Briglia (2020) found helping customers achieve their intended goals with the product or service purchased creates value for customers that increases customer loyalty outcomes such as retention, growth, and advocacy. The customer loyalty outcomes contribute to increased CLV. Popular retailer Nordstrom became well known for their exceptional customer service based on the attention to detail in the customer experience and the level to which employees were empowered to make decisions. The Nordstrom handbook's primary rule for employees is "use good judgment in all situations." The Nordstrom customer service stories are legendary, and many organizations have adopted some of the practices to increase customer satisfaction. Demonstrating a high degree of empathy for customers will provide clarity to see and solve problems that previously might never have been noticed.

Customer-obsessed culture is a highly touted mindset that Amazon and Microsoft leaders described as one of the key leading indicators for their organizations. Developing a customer-obsessed culture involves turning customers into fans who believe in, advocate for, and keep coming back to your business. One of Amazon's fourteen leadership principles is that leaders start with the customer and work backward. They work vigorously to earn and keep customer trust. Although leaders pay attention to competitors, they obsess over customers.

The idea of earning and keeping a customer's trust is highlighted by Covey (2008), who said that trust is a multiplier, trust is an accelerator, and trust is an energizer. Covey (2008) described the three vital reasons for trust as follows:

- Trust as a multiplier makes you better at everything you need to do. Low trust is a diminisher, a dilutor, a divisor, a tax.
- Trust as an accelerator enables you to do everything you need to do better and faster. In a VUCA world, speed is the name of the game today.

- Trust as an energizer energizes everything and everyone. It creates joy, fun, happiness, satisfaction, energy, excitement and safety for everything that you are doing. It engages the culture like nothing else.

The way to sustain and moonshot customer brand loyalty is to accelerate trust. The common practice used to measure if we are winning with brand loyalty is to combine CLV and NPS. NPS is calculated based on a single qualitative question: How likely is it that you would recommend our company, product, or service to a friend or colleague?

NPS is scored on a 0-10 scale and is categorized as follows:

- Customers who give a score of 9 or 10 are considered promoters. Promoters create greater value for the company because they tend to buy more, cost less to acquire and retain, stay longer, and generate positive word of mouth.
- Those giving a score of 7 or 8 are passives. Passives are satisfied, but much less enthusiastic customers. They are less likely to provide referrals and more likely to defect if something better comes along.
- Those giving a score from 0 to 6 are detractors. Detractors are more expensive to acquire and retain, defect faster, and can hurt the business through negative word of mouth.

Brand loyalty would be in good shape if customers demonstrate a high NPS rating and CLV retention of twenty-four months or more. This means that customers are loyal to our brand and willing to spend their hard earned dollars on our product and services.

SATISFIED CUSTOMERS FICTIONAL SHORT STORY

Six months had passed, and the proverbial first ninety-day strategy was now in place. Progress could be seen throughout Ashanti's organization. People were demonstrating behaviors that included happiness, purpose, autonomy, and mastery.

Ashanti invited her leadership team to acknowledge the success of leading the effort to increase happy contributing people and discuss the next ninety-day strategy. The new leadership consisted of Manny, serving as the value delivery champion; Tarasi, UX/UI champion; Basil, consultant leading coaching practice; Robert, a full-time employee partnering with Basil to establish the agile and coaching center of excellence; and Marisela, the data science champion.

Ashanti stood at the head of the sleek, glass-top business table as she looked at her team. She was proud of how far they had come, but she wanted to make even more progress with them. She knew that together, there was no end to what they could achieve.

"We have done great things over the last six months," Ashanti said, smiling at her team members. "Now, we must set our sights on building a customer-obsessed culture that will make our AR/VR customers fans of our products and help them to thrive in the metaverse. The approach to sustain satisfied customers is to ensure that we accelerate the rate of trust and pay attention to what the customer values." Ashanti paused. "By the way, we need a team name. Let's think of something fun."

Ashanti knew that injecting a bit of lightheartedness into the conversation could make things even better. She stood at the front of the room and wrote down the names the team came up with: 1) Git er Done, 2) Red Yarn, and 3) Product Slingers.

"We have one last vote," Ashanti said, having voted for Product Slingers herself. "What do you say, Marisela?"

"Hmmm..." Marisela hesitated. She considered going with Ashanti's favorite, but she had learned that in this group, she was welcome to have her own opinions. "Red Yarn," she decided. "It represents the connection or dependencies between features on a program board during big room planning."

"I love that," Ashanti beamed. "Now back to our product. Our bestselling product is the fantasy getaway vacation (FGV) offering a paradise escape that our customers enjoy very much. One of the big challenges faced by our customers, though, is the experience of motion sickness in our virtual environments. So how might we create an experience for our FGV customers so that they are comfortable and have limited motion sickness?"

"Well, virtual environment motion sickness is an industry issue and not just at AGR," Tarasi pointed out. "This is a known."

"I understand that" Ashanti replied. "But how might we create experiences that move beyond the limitations of industry norms?"

The sound of silence resonated in the air, leaving room for the distant sound of dogs barking to fill in the space. Manny went on mute, having to go take care of the noise, leaving the rest of the team to contemplate this issue.

"Well, the net promoter score received from customers was meh," Marisela replied. "We have not been able to move beyond the passive level that represents customers rating our services with a score of seven or eight. These customers are satisfied with our products and services but are not wild fans who tell others about us. This is concerning because we know the power of word-of-mouth advertising.

"Be aware that the NPS rating system only informs us about the customer's sentiment and does not provide the reason why they are in the meh state of mind. We will have to do additional research to learn more about what is the root cause of the problem. How might we increase our customer loyalty rating to a promoter level with a score of 9 or 10?"

"That's a great point," Ashanti said. "Do you have any insight into why this is the case or how we might change those ratings?"

"Well, the technology platform used for our flagship product is old," Marisela admitted. "So, we may not be able to scale to work with newer VR headsets in the near future. The approach to our platform modernization must be done with thoughtful considerations to be agnostic to technology platforms. So, the real question seems to be: how might we modernize our products to be agnostic to technology platforms?"

Ashanti nodded. She felt a little disheartened by this, but she wasn't about to give into those negative emotions. Instead, she was focused on finding solutions.

"It's obvious to me that we have several high priority items that must be considered to ensure that we get the biggest bang for the necessary investments," Ashanti said. "What I heard was 1) fix the motion sickness issues with our flagship product, 2) improve our customer loyalty challenges so they are no longer meh, and 3) modernize our technology platform. Is that correct?"

The team members' response was barely audible. Everyone sounded a little overwhelmed by this. But Ashanti was the team leader for a reason, and she wasn't going to let morale drop like this.

"Come on now," Ashanti said in a loud, animated voice. "I know you all have more energy than what you are showing. Our jobs are on the line here! The whole team needs your full commitment to solving these issues."

"There is a roadmap backlog that requires prioritization," Basil reminded them. "We can use the cost of delay to ensure that we begin to put into practice the customer-obsessed approach that Ashanti mentioned as a shift in our culture. The cost of delay is a simple arithmetic formula that sums three values. Let me share a few slides with the values."

Basil brought up a slide that read:

1. User-business value
2. Time criticality
3. Risk reduction/opportunity enablement (RR/OE)

The user-business value represents the relative business value to the user or business and the revenue impact to the business.

Time criticality defines how the user-business value decays over time. The time criticality considers the fixed deadlines that are constraints for customers and if customers will wait for the delivery of a specific feature or move to other competitors' offerings.

Risk reduction/opportunity enablement represents what else this feature will do for the user-business value. RR/OE informs us if other business opportunities can be amplified or there is a reduction in risks for future delivery.

"The CoD will help us make different decisions if we understand economically how much financial opportunity is lost," Basil said. "For example, what decisions would we make if the opportunity loss is $1,000, $10,000, or $100,000? Here is an illustration of the cost of delay computation in figure 6."

Cost of Delay (CoD) = User-business value + Time criticality + RR | OE value

***Figure** 6: Cost of Delay. Source: Reinertsen (2009)*

"An outside partner may be required to modernize our technology platform," Manny piped in. "Blending a cloud solution with the Unreal real-time 3D creation tool for photo-real visuals and immersive experiences are not skills that we currently have in our staff. We may also want to think about using VR-related anime as another option to connect with millennials and other customer segments."

"We have to ensure that we train our people to use agile practices to be able to deliver small increments of value," said Robert, leaning forward thoughtfully. "The last time we used the traditional approach to deliver the VR platform, it was a disaster.

"We did not have the opportunity to pivot when things weren't going well. It took us between nine and twelve months before we knew things were off the rails because the project manager continued to re-baseline the project schedule when things were not green. I would like to avoid using the red, amber, green (RAG) model to indicate the teams' progress this time."

"We heard many perspectives on the strategic work that will help us to reach the desired customer-obsessed culture," Ashanti said as she wrapped up the meeting. She made a list in front of them. "The three opportunities are: Fix the motion sickness issues with our flagship product, improve our customer loyalty challenges so they are no longer meh, and modernize our technology platform. Are we good with these options? Can we do this?"

"Yes, we can!" the team replied with enthusiasm, then erupted into loud laughter.

Ashanti shook her head, glad to have her enthusiastic team back. "You guys are so snarky," she laughed. "Basil, I will need your help to use the CoD method described earlier to prioritize the work. I would like to include the senior leadership team (SLT), our core team, and some of the beta customers in the CoD event. This approach would move us beyond the echo chambers of just our team members."

Ashanti reached out to her SLT peers to get on their calendars. It required a minimum of thirty days to schedule time with the SLT.

"This lack of availability is unbelievable," she muttered as she tried to schedule a meeting. Fortunately, she was able to find a calendar time slot thirty days out that aligned on the SLT calendars. The CoD event date was set for June 30.

Next on the list was to invite three beta customers to participate in the CoD event. The first customer had to be one of the large distributors that were involved with clients across the globe. The second customer was a small boutique that provided space for people to experience the AGR VR exotic vacations. Third on the list was a high-volume user who spent a lot of hours on their platform.

Ashanti made sure the plan was in motion as her team worked diligently to ensure success on their event date.

*****

On June 30, the optimism could be felt in the large, open meeting room. Everyone was lively and the Red Yarn team was ready to collaborate with the SLT and three selected customers.

The meeting was going to be a hybrid format that included some people in a collaboration room face to face and others connecting remotely.

The right-here-right-now (RHRN) small boutique owner, Celia, and three of her team members; the high-volume customer, Baylie; and some members of the SLT team were in the box in the virtual meeting room. Everyone else was in the airy, light, black and white meeting room at the office.

Anything can go wrong with hybrid events because there are two groups to be concerned with. First are the people in the room. The second group to engage are the people tuned in remotely. During COVID-19, the Red Yarn team members learned how to navigate the virtual world, but this was their first time being part of a hybrid event with so many high-ranking leaders.

Robert kicked off the meeting by saying, "We are going to start this meeting a little differently by using the core protocol check-in format to introduce ourselves. The format follows:

"1) 'My Name is …,' 2) tell us what you like about summer and why, and 3) say, 'I am in.' The rest of us listening will then say, 'Welcome,' in a loud and enthusiastic voice. Are you ready to try?"

This was followed by various nods and agreements through both platforms.

"I am going to model the introduction model and please don't forget to say, 'Welcome' after I say, 'I am in.' My name is Robert, leader of the agile and coaching center of excellence. I love summer because of the amazing fruits like mangos, nectarines, and strawberries. I am all in!"

"Welcome," the SLT, customers, and Red Yarn members said in soft, hesitant tones.

"That was disappointing," Robert said loudly. "I need you to use your outside voices with plenty of energy. You can do better than that!"

The attendees' energy and response volume picked up after being pushed by Robert as the second person checked into the collaboration space. There was laughter in the meeting spaces as the attendees began to provide more interesting answers about the things they liked about summer.

"I love summer because I can wear less clothes and feel less restricted," one of the customers said.

"I can relate," Celia agreed. "And who couldn't use a tropical vacation right now?"

There was more laughter and friendly comments. Robert could see that people were beginning to connect, and he knew this could allow for better collaboration during the hunt for selecting the highest priority initiative for investments.

"We have over one million customers globally and their sentiment about us is meh," Ashanti said, putting the meeting into swing. "The last NPS score was passive. This means that our customers are lukewarm about our products and are not actively fans who promote us. The good news is that we are still maintaining a CLV of eighteen months on average."

"Excuse me, Ashanti," Harry, the CFO, cut in. "Our target CLV was thirteen months. So, I would say that we have exceeded expectations."

"That's true," Ashanti admitted patiently. "Although we are exceeding targets, Harry, the reality is that some of the customers could go away with a small downturn in the economy or competitors owning more market shares. It could also mean that people are paying the subscription fee and not getting value from our products and services. Please allow me to describe the opportunities and challenges that we identified by using machine learning algorithms, customer interviews, and market research to learn more about their pain points."

"Of course," Harry nodded, impressed by her drive and ambition. "Please proceed."

"First, our customers are experiencing motion sickness in our popular fantasy vacation VR product," Ashanti explained. "We must outsmart our competitors and not accept this as standard for VR products. Being the best means that we should solve tough problems and make innovation a central part of who we are.

"Second, our customers' sentiment about our products and services is meh. This is heartbreaking news, but there is an opportunity to shift our customers' loyalty and make them fans.

"Third, our technology platforms are old, and we will not be able to work well in the metaverse. Our products need to work seamlessly on multiple platforms. I need your help with prioritizing the top three opportunities that include: fixing the motion sickness issues with the fantasy vacation VR product, improving our customer loyalty challenges so they are no longer meh, and modernizing our technology platform to work seamlessly on multiple platforms.

"Each one of these priorities will require a considerable investment to move the needle in our favor. I want us to identify the ranking of the opportunities as number one, two, and three. Allow me to introduce Basil, a consultant that we think of as family, who will introduce us to a technique to help us understand the cost of delay for the three opportunities and challenges. Robert, please lead us in this important conversation and collaboration."

"Allow me to introduce the cost of delay (CoD) concept that we will use to prioritize the three initiatives," Basil said.

"The CoD idea was introduced by Reinertsen in 2009, detailing the importance of evaluating priorities through the economics lens to accurately quantify the economics of delaying value to customers. The CoD mathematical calculation sums the user-business value, time criticality, and risk reduction and/or opportunity enablement values.

"We want to be customer centric in our approach, and that is one of the reasons that we invited actual customers to attend this collaboration session. When we think about selecting one opportunity over the other, there is a CoD associated with our choice. We must consider the impact of the CoD of one choice over the other. CoD enables us to evaluate each option's economic impact of a delay to the value delivery. The Scaled Agile Framework (SAFe) first principle is to 'take an economic view,' according to Reinertsen (2009), to deliver value early and often in order to learn fast."

As Basil paused, those involved in the conversation thought about how much they liked the idea of this approach. The analysis made sense to them, and they all appreciated how it was a customer-centric approach. The customers especially were grateful that they were taken into such consideration.

"Numeric values are needed to use the CoD calculation," Basil continued. "And the Fibonacci numbers of 1, 2, 3, 5, 8, and 13 will be used." Basil pulled up a document on the screen stating:

The user-business value context represents:

A. Revenue impact
B. Potential penalty or negative impact
C. Customer product or service preference

The time criticality context represents:

A. Fixed deadline

B. Wait for us or move to another solution

C. The effects on current customer satisfaction

The risk reduction and/or opportunity context represents:

A. The risk of the present or future delivery

B. Value is available in the information we will receive

C. Enable new business opportunities

Members of the SLT and selected customers were given digital copies of the CoD context along with six the Fibonacci numbers of 1, 2, 3, 5, 8, and 13 that would help with ranking the importance of each CoD value.

Tarasi and Marisela facilitated the CoD ranking activity to establish the priority of the initiatives based on CoD to deliver value. Marisela led the user-business value conversation and began by sharing the context for the user-business value.

"Was the user-business value context clear?" Marisela asked.

"What was meant by potential penalty or negative impact and to whom does it apply?" asked Celia, the owner of the boutique.

"That is an excellent question!" Marisela answered. "This may apply to AGR or to your organization. The penalty could be driven by regulatory constraints, and the negative impact could be a delay in delivering one of the three described initiatives."

"Thank you," Celia replied, grateful to be included in the type of company discussions that consumers are all too often left out of.

"You're welcome," Marisela replied warmly. "It's our goal to make sure everyone is on the same page today. Now, I'm going to continue, but I'm asking you all to review the six digital cards that contain the Fibonacci numbers of 1, 2, 3, 5, 8, and 13. On the cards, 1 represents the lowest value and 13 is the highest.

"When asked, you will place one of the six digi-cards on the digital collaboration board. Furthermore, we would like you to use 'yes and …' or 'how might we …' when responding with an alternative opinion to one of the participants' ideas. Tarasi and I will model what that would sound and look like. My user-business value number is a 3 because revenue is not the issue."

"I will go next," Tarasi replied. "My user-business value number is an 8; yes and revenue is critical to invest in shifting customer loyalty."

"Was the 'yes and …' a knock off from improv?" asked Harry, smiling at its use. "I remember taking a course and using that technique several years ago. It's a clever and useful way to approach things.

"I will go first, and here is my 8 that represents revenue impact. If there is not a shift in customer loyalty, revenue will be affected sooner or later. Having a maa-aay ranking does not bode well for our company."

Laughter filled the room and virtual space as the tension eased at Harry's friendly and open demeanor. People began to imitate Harry by using maa-aay in their own responses.

"I am placing a 13 for customer product or service preference," Baylie jumped in. "So, yes and as a customer, my concern is about satisfying my job requirements based on the utility that I receive from the fantasy getaway vacation offering. I do spend a lot of time in FGV to reduce the high stress of my work each day, sometimes multiple times during the day."

The collaborative conversation continued, and the group aligned by assigning a user-business value of 8. Second was the time criticality context that represents a) fixed deadline, b) wait for us or move to another solution, and c) the effect on current customer satisfaction. Third is the risk reduction and/or opportunity context that represents a) the risk of the present or future delivery, b) value is available in the information we will receive, and c) enable new business opportunity.

After hours of rich conversations, the CoD ranking activity results were:

1. Meh customer loyalty is a 16
2. VR motion sickness 13
3. Technology modernization 11

"Although we used the CoD to determine the priority and sequence of the initiative's strategic direction," Marisela said, "as we engage the team members responsible for development, the effort to build the initiative will be considered and may shift priorities. We will engage this group again to help with the decision to stay the course with the prior decision or pivot based on the new information and reprioritize accordingly."

"We achieved the objectives today by receiving your contribution to prioritize the strategic direction for the year," Ashanti said. "Thank you for giving so generously and helping us to begin the customer-obsessed culture journey. Red Yarn team, great work facilitating today's conversation by making it engaging and fun. Before closing our session with a retrospection, do you have any compelling questions that you need answered?"

There was silence in the room. "I for one would like to say that I enjoyed the session and appreciated being included," Harry said. The rest of the team nodded their agreement.

"Thank you, Harry," Ashanti replied. "If there are no questions, I will turn it over to Robert to facilitate the retrospective activity."

"Thank you for attending and contributing to the strategic priority activities to help us delight our customers," Robert said. "The purpose for the retrospection is to help us improve by sharing your experiences today.

"We will focus on three questions, and they are what went well, what did not go well, and what you would change. I ask that you create one response per digi-card. You may add more than one digi-card for each question. Let's begin by answering the first question of what went well."

After giving the group time to discuss this among themselves and come up with solid answers, Robert asked if anyone would like to share their answers. Celia was called upon first.

"I think one thing that went well was everyone was included in this discussion," Celia replied. "As a business owner, it's nice to be taken into consideration. I felt like my voice was heard, and I really appreciate that."

"I agree," Baylie responded. "I like the fact that the problems were brought to light and strides were made toward finding those solutions. I feel like the company takes our input seriously and will use it to better the products in the future."

There were nods around the room as they all agreed that this was a great aspect of the meeting. More things that went well were passed around. It seemed like the meeting had been a success.

"Wonderful!" Robert said. "Now, let's move to what did not go well."

That topic was a little more difficult for the attendees to speak up about. But finally, Marisela spoke up.

"I guess I would've hoped to always find more solid answers to solving these problems," Marisela admitted. "Though I suspect that's just the perfectionist within me. I always want to be doing more to make the company succeed."

Again, attendees agreed with Marisela. A few more complaints were brought up. But there were far fewer complaints compared to the positive remarks.

"That speaks to what could be done better," Ashanti said. "And I think in the future, we will do better at coming up with more solutions to make our company even stronger."

"That's what we're striving for," Robert agreed.

The conversation continued until everyone had the chance to share what was on their mind, and the meeting came to an end.

"Thank you all for coming here today," Robert said, smiling at all the attendees. "I appreciate how engaging the group was. And I know that together, we will make this business an even greater success than it has been. I am optimistic for the future, knowing we have such an incredible team to work with."

SATISFIED CUSTOMERS - INFLUENTIAL LEADERS' VOICES

Diana Larsen, author of *Agile Retrospective* and co-founder of Agile Fluency Model

I'm thinking about how the fluency model also identifies a path to delighting customers as well. Because I think that's a significant aspect for companies as they grow, even individuals. How can the agile fluency model help get to a state of satisfying our customers?

Right. The idea of the fluency model is to bring the team closer to their customers. Every zone has a different way of being close to the customer. Whereas in our zone we have a little area we call pre-agile. Which is when people are not working as a team; they're still individual contributors. They're getting their tasks assigned by their managers. They're working in isolation; maybe they do some pairing, but they're not pairing all the time.

There are still a lot of companies that have that as their work process model. Those folks tend to end up building in that componentized way. "I only work on the back end. I only work on the front end. I only work on the database connections. I only work on the ..." whatever it might be, rather than thinking about the whole product or the whole that is needed. That shift to the focusing zone is very much about a shift from working in that way in terms of, "I built this kind of thing, and I built it in a very cool way. I get to use a programming language that I particularly like," to now, "I'm working with my teammates and each thing we build, we have a connection that tells us this is something that's important to our customer. This is a whole piece that will enable our customer. This feature, or part of a feature, will enable our customer to do something they weren't able to do before." So, in the focusing zone, they have that communication through business liaison. In the Scrum agile practice, that would be a product owner. It could be different business analysts sometimes; even quality folks will be giving that information about what the customer likes about this, what they don't like about this. So, there's more awareness of the customer there.

Then, in the delivering zone, it's not only awareness of things the customer wants this product to do, but there's an awareness of how often they need to get it, how often they need it updated. There's more of a cadence. We learn more about our customer because now we know not only what they want, but how often do they want it and how important is the long-term maintainability and sustainability of this through user experience (UX) and continuous integration and continuous delivery (DevOps) and the other partners that come into play in delivering.

Then, in the optimizing zone, the team is taking on the whole product. Now they really need to understand their customer. Not only do they need to understand their customer's current needs, but they also need to understand their customer well enough to be able to anticipate how new technologies might be useful to this customer in ways the customer hasn't even thought about yet. But we understand the nature of their situation and their problems well enough that we can see that this could be helpful, and we can begin to start talking to them about that and building that into the product even before they're asking for it.

So, it's an increasing closeness to the customer and an increase of bringing that business perspective into the team as you go through the zones. So, by the time you get to optimizing, the businessperson—whether that's a product management or product development and marketing person—whoever that might be is right in the team all the time as opposed to coming in and out and bringing information, which is useful in the focusing zone. But we just keep getting closer and closer. That's how the customer fits.

One of our licensed facilitators and dear friend and colleague, Ellen Gottesdiener, of course has done a ton of work in the customer and product management space. Her book is *Discover to Deliver*, which includes all the product management stuff. She has begun to do some work on a product management or product development fluency model because we see that there is a parallel path, but we don't know exactly how to describe it yet. So, she's been working on that because it is so important. Customer collaboration over contract negotiations, right out of them. It's a quarter of the manifesto. It's right there.

So anytime we're talking about agile adoption or talking about agile practices, we must keep in mind we are also talking about what is our degree of closeness to our customer, and how do we make it tighter.

Howard Sublett, former CEO, Scrum Alliance

By any definition, customers are satisfied when they receive value in excess of the amount they paid or the effort they provided. In other words, what you've given them is worth more than they have invested.

Marty Nelson, Alchemy Code Lab founder

Coming from a software development background, I usually think of the users of our products and owners of those product systems. Are they getting value in a good way? And sometimes that means staying out of the way! I had a manager early in my career who likened the role of software developer to a sports referee, "You're doing your job when nobody notices." With more consumer-facing apps and growth of technology, there's more opportunity to make a splash, but I think software that just quietly does the thing it is supposed to do is still underrated.

These days, I spend time designing apps for clients, and we always ask, "What's the simplest happy story we could tell?" If you can't find the emotional validation, I think it's hard to satisfy users or the customer.

We work with entrepreneurs and often the first MVP is about learning and movement and direction. That client needs to feel that what we're doing is moving them toward the goal. It's not a one and done viral explosion. For example, one of our clients is exploring new subscription pricing models for real estate. We built the initial app for just him, and he had some success but realized he needed to think about getting multiple Realtors onto the product as a platform. He was able to secure a job at a realty firm that allowed him to manage a large group of realtors, and we moved into a second phase to make the product multi-platform so he could bring those Realtors on board. This is an exploration into an unknown problem, and he's been very satisfied with how we have worked alongside him to create opportunity.

With the school and getting people into good software developers, you'd think it would be easy to have satisfied customers. Sometimes hard, challenging things that force you to grow can be rough. This aspect of the business feels a lot like my parenting experiences in life. You push people to grow and put in the work. It pays off, but sometimes there's this lingering resentment because it was emotionally difficult. Or conversely, they can be dismissive of the help you gave. Usually people come around, just like our teenagers do when they become older adults. I've seen this same thing in business environments where a new internal software product disrupts people's routine and sense of purpose, but as it gets adopted and the benefits become clear, you'll see more satisfaction.

Dave West, CEO, Scrum.org

There are normally two types of customers. One customer is the end user of the thing you create. The other is the company or organization that is funding that thing to be delivered. Those two groups are satisfied in very different ways.

End users are satisfied by products doing what they need, wowing them by not wowing them. For example, Amazon Prime gave me an out of the box, simple solution to a recurring annoyance of paying for shipping and wanting stuff yesterday. It did it in a way that just felt natural. But if you had asked me what I wanted, I am not sure I would have said Amazon Prime! Amazon Prime wow'd me. It filled a need I did not know I had.

Funders are also more complex. Traditional project management would say all the sponsor wants is "on time, on budget delivered what they asked for." But for people funding work, that is a really hard question to answer—the what. That means that increasingly funders want to see customer benefits realized. For example, they want to see NPR increased or support tickets decrease. Now, if you dig into the real motivation for their change of heart, it is often associated with the objectives of their boss or their boss's boss. These people have come to realize the green project dashboard provided little real knowledge on the investment they made. So, they provide guidance to the people involved in funding to "make it more strategic and measure more customer stuff." This results in a combination of traditional project metrics and more modern, less well-understood approaches.

From my experience, the best way of keeping both parties satisfied is to provide frequent, observable progress allowing them to provide feedback. To ask them why they like something or don't like something else. To use techniques like 5 Whys and Fishbone to evaluate their feedback. To ask lots of questions and be humble. The bottom line is we need everyone to know we are on a journey together, learning and improving as we go.

BUILDING A CUSTOMER-OBSESSED CULTURE

The customer-obsessed culture indicates that it is important to give customers a compelling series of experiences to keep them coming back for more and not limit the experience to a single satisfying experience. The customer-obsessed culture is documented and shared by retailer Nordstrom, software giant Microsoft, and online retailer and technology platform giant Amazon. The emphasis is to do what is necessary to retain and attract satisfied customers. Building a customer-obsessed culture requires discipline and a will to run limited experiments to learn about customer habits and needs. This culture is not perfect and will require going outside of the status quo to spot customers' problems and aspirations.

The customers' jobs to be done represent what products or services a customer will hire and fire to complete a job they need done (Christensen, 1997). For example, fast food restaurants satisfy the need for quick food at a reasonable price that is tasty. At any time, the customer can fire one fast food restaurant and hire another. When the business owner or entrepreneur understands the customers' jobs to be done, the chance of success is good until a competitor enters the market. The customer does not exist to buy your product or services. You exist to service the customer's job to be done.

Customer jobs can be categorized into three types of jobs that are functional, social, and emotional:

- Functional jobs are practical considerations that are usually straightforward.

- Social jobs are driven by a sense of duty, optimism, aspiration—the things we feel we must do as part of our role in relationships and society.
- Emotional jobs come from our fears, preferences, likes and dislikes, insecurities, and indulgences.

An example of a functional customer job to be done is that when working from home during COVID-19, I need a primary reliable internet connection and a secondary option in case of failure. A social customer job to be done is underestimated students need Future of Work workforce development capabilities. An emotional customer job to be done is I want an eco-friendly car that makes me feel like I am doing something about climate change.

When we want to understand a unique value proposition for customers, first start with learning about the customers' jobs to be done by having conversations with the customer about what job(s) the customer can complete with your product or service. Second, identify the pain relievers for the customer by discovering what problems the customer wants eliminated when using your product or service. Third, examine the gain creators of your product or services that will benefit the customer when using your product or service. The unique value proposition canvas is illustrated in Figure 7.

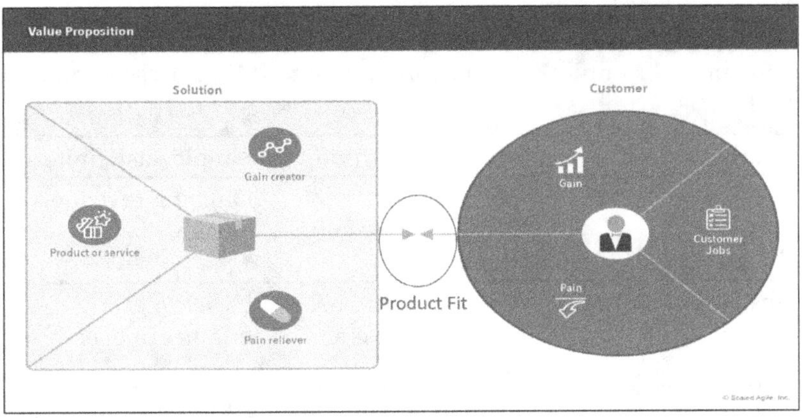

***Figure 7**: Unique value proposition. **Adopted from**: Osterwalder, A., Pigneur, Y., Bernarda, G., Smith, A. (2014)*

The potential customers, also known as prospects, require a thoughtful approach to obtain insights into customer jobs, customer gains, and customer pains. Table 1 describes examples of thriving businesses with a value proposition to satisfy customers' unmet needs.

***Table 1**: Customer Value Propositions*

Business Examples	Customer Benefit (Value)	Business Benefit (Value)
Starbucks	Simple, fast, finite	Simple, fast, finite
Netflix	Set-it-and-forget-it, less time spent on transactions	Ongoing revenue stream
Instagram	Low barrier to use, no charge to end users	Easier to grow number of users
New York Times Online	Trial period at "no cost"	Hook customers without charging, easier to grow number of customers
Metromile Toll Service	Don't pay for something you don't use	Appeals to those who have relatively low usage rates or big variations in usage

The following are some questions to help you define customer pain relievers:

- Does your product/service provide savings? (Regarding time, money, efforts, etc.)
- Does it improve the emotional state of your customer? (Decrease grief, irritation, things that bring a headache)
- Does it fix the defects of existing solutions?
- Does it remove the difficulties or problems that your customer faces?
- Does your product/service exclude the negative social consequences that are encountered or which your customers are afraid of?
- Does it reduce the risks that your clients are afraid of?
- Does it help your customers sleep better at night?

- Does it limit or eradicate common mistakes that customers tolerate?
- Does it eliminate barriers that keep your customers from implementing the specific solution?

The following are some questions to help you define customer gains creators:

- Does your product provide savings that make your client happy?
- Does it ensure the results that the customer expects?
- Does it simplify the work or life of the customer?
- Does it give something that your customer wants to get?
- Does your product/service reflect some of the dreams of your customer?
- Does it give positive results that meet the criteria for success and failure of the customer?

By gaining insights into the customer jobs to be done, pains, and gains, the real possibility of defining a fit between the innovation you created (including products or services), pain relievers, and gain creators. The opportunities to support satisfied customers with new product features will delight customers.

Focus on the Customer's Unmet Needs

To understand how to retain and enable satisfied customers, first understand the customer's unmet needs. Identifying the customer's unmet need is the first principle of innovative entrepreneurial organizations. The second principle is fulfilling the customer's unmet needs. In the hybrid world of conventional and digital customers, discovering customers' unmet needs requires expanding how to find customers' problems and aspirations. Figure 8 illustrates the four ways of looking to diversify the approach to spot customers' problems and aspirations.

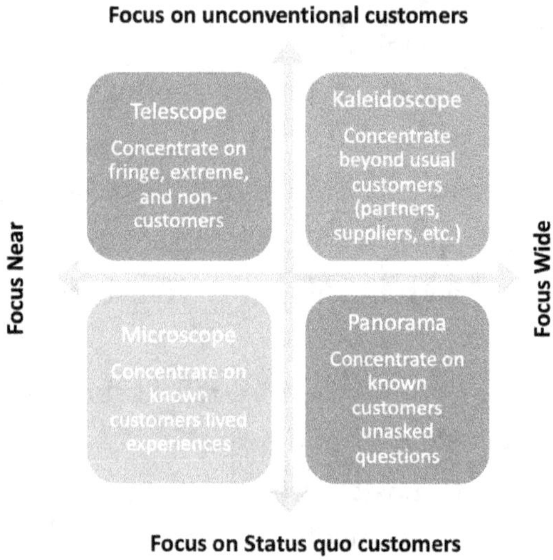

***Figure 8**: Spotting customers' problems and aspirations.* ***Adopted from**: Barsoux, J., Wade, M., and Bouquet, C. (2022)*

How might a pet parent retailer use the lens of spotting the customer's problems and aspirations? A scenario from one of my clients was to reduce friction in the online and in-store checkout process. The scenario can be explored through the use of the telescope, kaleidoscope, microscope, and panorama strategies to understand customers' unmet needs. Use the telescope strategy to concentrate on fringe, extreme, and non-customers by observing customer behaviors in the online and brick-and-mortar stores. What products did they browse and not purchase? Apply the kaleidoscope strategy to concentrate beyond the usual customers to include partners, suppliers, and other entities inside and outside your sphere of influence by holding focus groups with people browsing your stores who provided contact information through a sales funnel. A sales funnel is a digital tool that seeks to learn about non-customers by capturing name, email, and phone number to begin a digital conversation.

Employ the microscope strategy to concentrate on known customers' lived experiences by using micro-surveys to capture immediate customer feedback. This is the common starting point to address overlooked opportunities that can be addressed. Exploit the panorama strategy to concentrate on known customers' unasked questions through aggregated data, which includes errors, complaints, and accidents, that amplifies unserved areas. The panorama strategy would use data stored in knowledge management systems to provide insights about customers and customer service agents' feedback. Knowledge management systems may include emails, customer service systems, and word of mouth.

Artificial intelligence (AI) digital tools increase the ability to implement the telescope, kaleidoscope, microscope, and panorama strategies to discover the customer's unmet needs. AI digital tools support the ability to scale and include the four customer unmet needs strategies by using AI algorithms to explore large numbers of customer behaviors in pseudo real time. The AI digital tools approach provides the options to study extreme, fringe, and non-users to uncover pain points that are important to the standard customers. Hard-to-reach customers that include outlier prospects and challenged customers can be observed through digital communities that people use for professional development, friends and family engagement, and consuming information about products and services. This means that social-listening tools, unstructured data-scraping algorithms, and semantic AI tools support quickly sifting through large data populations to identify patterns that help to discover unmet customer needs.

With the scalable AI digital tools come highly publicized ethical risks. The two most common are invasions of privacy and unintended bias. The risks may cause reputational, regulatory, financial, and legal threats, leading to public distrust. Identifying and planning the ethical risks mitigation must be included in every initiative execution to eliminate waste of valuable organization resources of time and money.

Cognitive biases are predictable patterns of thoughts and behaviors that lead you to draw incorrect conclusions.

– David McRaney

Anchor on Empathy for Your Customers

Enough cannot be said about anchoring your customer relationship by expressing empathy for their needs and interaction with your products, services, and people. Equipping happy contributing people with the power of empathy would increase the level of customer satisfaction. Figure 9 illustrates the visuals and inquiries that increase awareness of customers' emotional journey with your product or services.

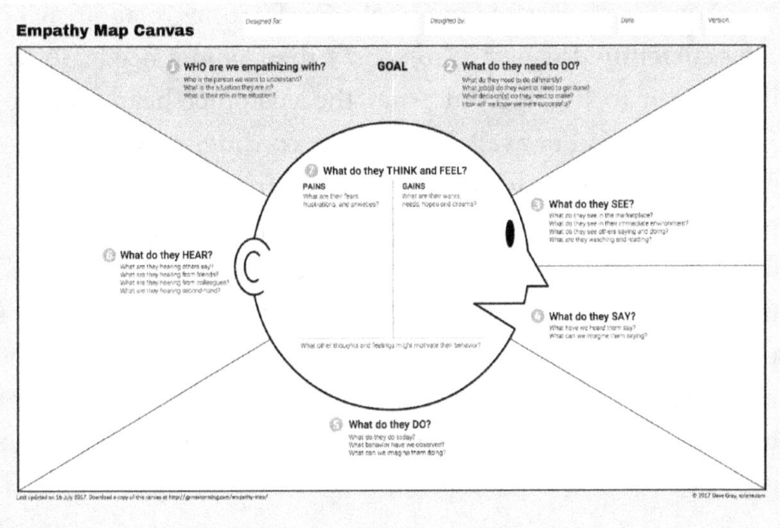

***Figure 9**: Empathy Map Canvas. **Adopted from**: Gray, D., Brown, S., and Macanufo, J. (2010)*

Following the empathy canvas map from step 1 through step 7 greatly improves the insight into what can be imagined of walking in the customer's shoes. Understanding the unmet customer needs begins with learning more about the customer goals, highlighted in steps 1 and 2. Begin the empathy journey by learning about the customer, the situation they are in, and the role played in engaging with the organization's products or services and happy contributing people. The knowledge of what the customer needs to do is gained by observing or inquiring about what needs to be done differently from the existing experiences or decisions for satisfaction. Understanding what the customer sees (step 3) in the marketplace, in their immediate environment, other people doing, and as available information gives us an idea of the mental picture the customer has developed. Listening to the customer (step 4) to hear what is said about the experience with existing products creates another piece of the puzzle.

Another observable moment is to learn more about what they currently do (step 5) to satisfy their unmet needs. In the interaction to learn about what the customer hears (step 6) from the myriad of available sources, consider how customers are influenced by information received from social media and interacting with people in their circle of influence. The most difficult part is to understand the customer's feelings and thoughts (step 7). However, understanding the customer's pains experienced and gains desired will move you in the right direction. Investing time to demonstrate empathy for the customer's pains and gains may build trust and increase a continued business relationship.

Customers' Response to New Innovations

All customers may not readily adopt the new innovations that are brought to market. The customer response to new innovations in relation to market share is illustrated in Figure 10. The customer behavior patterns demonstrate the adoption of new innovations in the market in contrast to customer adoption. Fleiter and Plötz (2013) found "the adoption rate, measured as the share of new users in a given time interval compared to all potential adopters, increases continuously until it reaches a maximum at the point of inflection (of the cumulative number of adopters)."

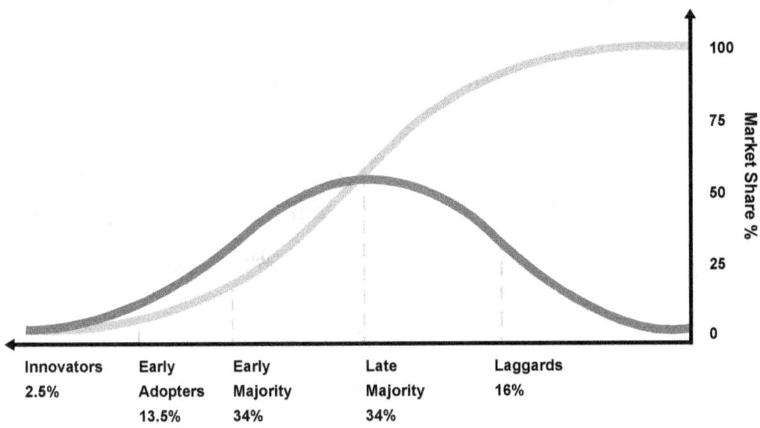

***Figure 10**: Innovation hype cycle and s-shaped adoption **Adopted from**: Fleiter, T. and Plötz, P. (2013)*

Moore (2006) noted that many technology products face a "chasm" between the expectations and requirements of early adopters and the rest of the market. Figure 11 describes the chasm between maturing a product or service to connect with the early majority or pragmatists group.

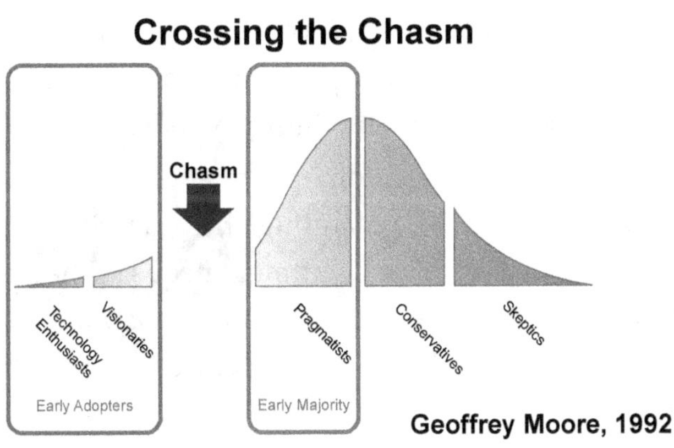

***Figure 11**: Crossing the chasm **Adopted from**: Moore, G. (2006)*

There are five categories of prospects (potential customers) described by Moore (2006). Your potential customers' response to new innovations may fit into one or more of the following five categories:

1. *Innovators/Technology Enthusiasts* - seek out novel technology; it's like a hobby for them. They are limited in numbers. Other people see them use new things and feel braver about trying them.
2. *Early Adopters/Visionaries* - looking for breakthrough technology, and they are willing to pay well to be first with the new technology. This group relies on its own intuition and vision to make buying decisions.
3. *Early Majority/Pragmatists* - practical-minded consumers. If a product seems useful, this group will try it. The Early Majority are cautious of fads.
4. *Late Majority/Conservatives* - wait for something to become well-established. They don't feel confident in their ability to deal with technology and often buy from big companies.
5. *Laggards/Skeptics* - for personal and/or economic reasons, they are not looking to buy new technology. Existing or potential customers may be laggards because they are not willing to take risks.

Awareness of the customer's rate of adoption to new innovations helps with planning the activities necessary to explore the customer's unmet needs. The telescope, kaleidoscope, microscope, and panorama strategies and AI digital tools provide the insights to discover the customer's unmet needs.

Finding a fit between an organization's innovation and the customer jobs to be done is complex, time consuming, and costly. Understanding the customer's unmet needs is most important to protecting the organization's vitality. By starting with customer empathy, critical insights are obtained about the customer. It is all about how much the customer desires to purchase your product or services.

The use of AI digital tools to gain insights about customer patterns may reduce the time to produce an outcome that is useful. Be aware that AI algorithms are known to create costly risks based on racial, gender, and other biases. The complexity and cost may still be high. The end goal is simply to create a community of satisfied customers who are loyal fans.

SUMMARY

Understanding the customer's jobs to be done, pain points, and desired gains is part of creating a transformation business that identifies barriers, evaluates the customer's journey, and translates the new knowledge into an offering that satisfies the customer's unmet needs. The customer-obsessed culture is real and will demand the use of every available tool and practice to gain insights that allow the development of a long-term partnership with customers. The full-fledged customer partnership will enable the organization to create an economic value that improves a fundamental aspect of the customer's life and, in turn, creates a new context for the organization that did not exist prior. The customer obsession is a focus on the customer's success in achieving their desired results.

Thriving Business

The people who are crazy enough to think they can change the world are the ones who do.

— Steve Jobs

Agility is an organization's ability to harness changes for its competitive advantage.

— Ken Schwaber

INTRODUCTION

A thriving business requires the lens of being customer centric and building teams around delivering value. Satisfied customers and happy contributing people are the epicenter of enabling a business to thrive, especially in a volatile, uncertain, complex, and ambiguous (VUCA) environment. Stiglitz (2020) emphasized the importance of organizations focusing on maximizing value to customers, employees, and the broader community as well as shareholders. The people who deliver value fastest and the most frequently are the likely winners of satisfying customers and helping organizations to thrive. This is not to say that there will not be one-hit wonders that come along and provide value. Sir Isaac Newton's first law of motion taught that an object at rest tends to stay at rest and an object in motion tends to continue in the direction in which it is currently moving. The same applies to delivering value. The organization with a focus on delivering value frequently will continue in that direction, focusing on unmet customer needs by developing a customer-obsessed culture.

After watching "The Fourth Industrial Revolution" video many times, I realized that the message being conveyed is that we live in a globally shifting business environment that is shaped by change, competition, connectedness, and complexity (Schwab, 2016). Similarly, the term VUCA shouts that we are in a space of craziness (Bennett and Lemoine, 2014). Customers are seeking value to satisfy their needs with access to multiple providers. Businesses are faced with competition seeking to be disruptors by providing value at a lower price point. Christensen (1997) pointed to this disruption phenomena.

With easy access to inexpensive technology and the ability to connect with customers, the opportunity for disruption is easier than ever. Customers are seeking the best value at the lowest price point, but in some cases, they are willing to pay the premium rates for a status-elevating product. An example of a status-elevating product is the mobile phone when it was first introduced in the 1990s, then the Apple iPhone in the following years.

In the last twenty years, I would say that we have been in a heightened state of accelerating innovation. Irwin (2016) stated, "We are in a golden era of innovation in which digital technology is transforming the underpinnings of human existence." The ability for people to collaborate across the world is unprecedented. The distance between us is shorter because of the ability to see and speak to each other through the use of technology. The application of digital transformation in medicine, transportation, and education has given answers to questions that were previously impossible. We can now run multiple experiments at a low cost and evaluate multiple results to help decide the appropriate actions going forward by using cloud technologies and AI.

Transformation is a popular buzzword, and companies are trying to understand what they need to do to remain competitive while balancing being able to sustain the continuity of running the business day to day and building for the future. Transformation is reinventing the organization by holding the leadership team accountable for engaging with the question of how best to shape the future by co-creating with the organization's happy contributing people and satisfied customers. Transformation is a shift in ways of working, culture, and leadership behaviors. More organizations are using digitization to transform ways of working and business operations, even companies that need brick and mortar locations. During a mobile world conference, Satya Nadella, Microsoft CEO, was quoted as saying that every company is now a software company (Holmes, 2019). Is Starbucks a software company with the massive amount of digitization that makes it frictionless to order and fetch my favorite morning brew without waiting in line? My experience tells me that Starbucks' goal is a customer experience that delights the customer no matter the in-store or digital interaction.

As organizations launch into the transformation journey, two questions are important to ask. First, is the transformation driven by internal needs or external forces? Second, does it need to happen quickly, or do you have more time to transform? There are several transformation classifications, but this segment of the book will use the adaptive transformation that is initiated internally with a desire to go fast by increasing transparency, collaboration, and innovation. The other is disruptive transformations that are driven by external factors like regulatory demands, market conditions, global conflicts, and others that often require a quick response.

Adaptive transformation is the response to internal needs that aims to change the status quo of ways of working, mindsets, and leadership. Yesterday's approach to innovation is not today's model for responding to customers' unmet needs. The approach of leaders telling people what to do is passe and limits the ability to develop happy contributing people that enable satisfied customers. Currently, the most popular adaptive transformation is the lean-agile mindset with frameworks consisting of engagement and operating models that provide common language and approaches for enabling customer-centric partnerships. Engagement models are customer-centric activities that support users' (internal) and/or customers' demands to request a service or product. Operating models are practices and ways of working that deliver value to users and/or customers. Figure 12 illustrates the digital operating model to deliver value now and in the future.

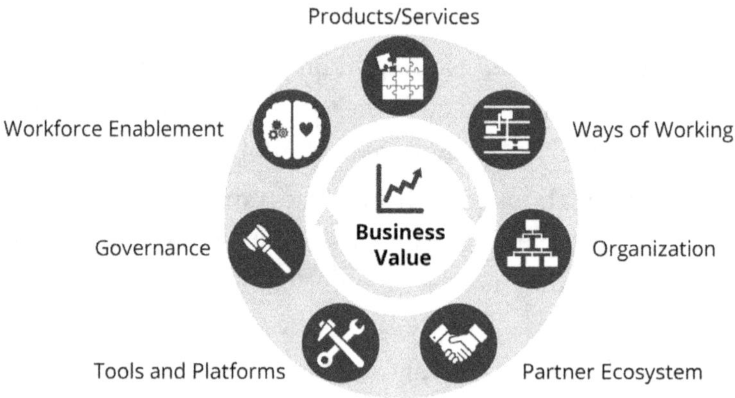

***Figure 12**: Business Value delivery model **Adopted from**: Mäkinen, Adam, and Ko (2022).*

Business value is the key theme and central to a thriving business based on strategic themes and business objectives. Products or services are customer-centric portfolios that drive value through revenue, customer delight, and market share. The adaptive ways of working represent lightweight practices that are adaptive to speed as markets, technology shifts, and business objectives change over time. The organization architecture enables a business- and customer-centric culture that includes a business-aligned and empowered agile team composed of T-shaped roles and skills. The partner alliances are the internal and external business partners that augment gaps in existing capabilities. Tools and platforms are frictionless technologies that enable collaboration, provide insights into opportunities and threats, and increase automation. Governance provides the appropriate controls, performance visibility, legitimized shared accountabilities, and clear decision rights. Workforce engagement/development is worker-driven continual improvement and accountability.

The adaptive transformation employs business agility (BA) to enact shifts in the entire enterprise and not just IT. Business agility draws our attention to leaders, people, and operations, remembering that organizations are run by people for people. Traditional areas—including C-Suite, HR, marketing, sales, finance, supply chain, and business operations—are all in it. Leaders must champion agility in all business initiatives. Disruptive transformation is building the awareness and resilience to adaptively respond to opportunities and threats that originate externally to the organization. This includes wars in countries that affect energy and supply chains as well as regulatory laws that govern various aspects of the business, which may include the European Union (EU) General Data Protection Regulation (GDPR), Sarbane-Oxley (SOX), and ISO-9001. Inflation is another external factor that affects the cost of borrowing money to run and expand business operations.

A thriving business demands a leadership imperative that mandates the embrace of transformation that includes upskilling human capabilities, digitization, and servant leadership. There must be a clear strategy to democratize transformation across business units to build the organizational architecture that is nimble and easy to adapt to disruptions. You should deploy an adaptive governance practice that provides a clear line of sight of inflight initiatives to align and integrate where possible.

GENERATIVE AGILE LEADERSHIP (GAL)

The future of work leadership imperative demands a new way to co-create a collaborative working environment in partnership with direct reports, peers, and partners to foster creating emergent leaders. In reality, the trend started more than ten years ago with the call to develop more leaders and fewer followers. The servant and catalyst leadership styles became ideals for people to aspire toward and pursue. GAL is a new path toward a leadership practice that combines continual learning and performance-oriented focus to help the organization to create potential new sources of growth to thrive.

Generative culture gives birth to organizations and human beings that demonstrate the ability to achieve amazing outcomes during the VUCA times or normal times. Developing generative business culture is fundamental to frictionless operations and effectively collaborating with people to succeed in a highly competitive world. What can be expected in a generative culture? An environment where people learn quickly and innovate to meet the demands of customers and markets. People collaborate more effectively with a higher level of trust in all directions (across, up, down) of the organizational structure. The petty personal issues are dampened to elevate the goal of the products and services that are needed to sustain the business's competitiveness.

If you have worked in a safe, high-performing team environment, you know the emotional energy is off the charts.

It is intoxicating at times, and when you fail, there is a whole team focus to learn together from experiments. The blame game goes away, and we settle in a good place that states "we are all learning." Generative cultures also enable diversity, equity, inclusion, and belonging. You will find generative leadership that creates space for development and performance growth for the people working in the organization and community. The generative environment's generosity extends beyond the doors of the organization and into the global community.

GAL is based on Westrum's (2004) research on three organizational cultures: pathological (power oriented), bureaucratic (rule oriented), and generative (performance oriented). The GAL value system combines the continual learning and performance-oriented context to create a continual learning and high-performing organization. More specifically, it develops an environment that enables continual learning and high-performing teams. People's natural impulse to learn is unleashed when they are engaged in an initiative where they find a purpose that is worthy of their fullest commitment. Senge (1990) expressed that a generative learning culture requires new ways of looking at the VUCA world, whether in understanding customers or in understanding how to better lead a business. The generative learning culture expands the ability to see the systems that control events. Seeing systemic sources of problems, we eliminate the underlying root causes rather than focusing on the symptoms (Senge, 1990). An example of a symptom may include people not collaborating because they are on different teams. The root cause here is the silo team composition design and the lack of leaders setting the context of collaboration.

As discussed in previous sections, the GAL values include the following:

1. High cooperation – Focus on we
2. Messengers enabled – We trust you to achieve our goals
3. Risks are shared – Win and lose as a team
4. Bridging encouraged – Partner and have conversations to learn more
5. Failures lead to inquiry – Ask, "What did we learn?"
6. Novelty implemented – Run the experiment and evaluate the hypothesis
7. Abundance mindset – We have capacity to share with others

The GAL values 1 through 6 are derived from the work of Westrum's (2004) typology of organizational culture. GAL value 7 represents the mindset required to share knowledge and success with others in the organization. The abundance mindset is when a person believes there are enough resources, goodwill, and successes to share with others. Covey (2013) shared that the concept of scarcity mentality refers to people seeing life as a finite pie, so that if one person takes a big piece, that leaves less for everyone else. The scarcity mentality is the opposite of the abundance mindset.

The GAL practice follows the principle that generative leadership is a higher form (Senge, 1990) of leadership. Taylor (2016) found that generative leadership fosters innovation, organizational agility, and high performance over time. Generative leadership emphasizes the relationships between leaders and team members. Figure 13 illustrates the focus toward building organizations as communities for learning.

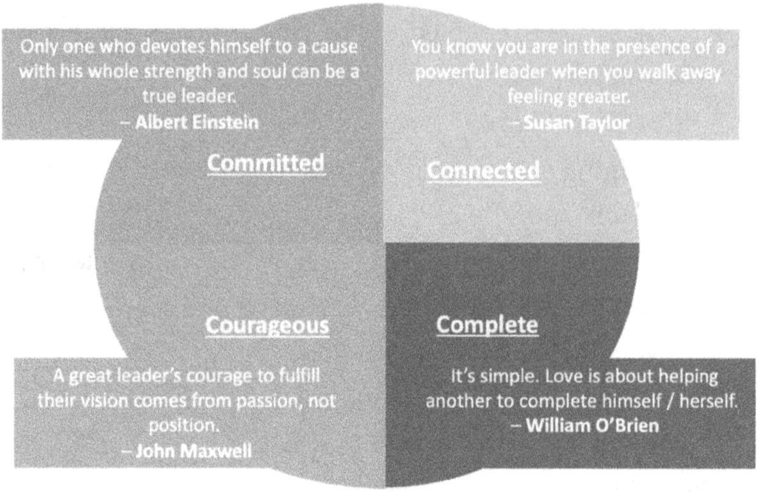

***Figure 13**: 4 Cs of Generative Leadership* ***Adopted from**: Taylor (2016).*

The committed GAL advocate has purpose and also partners with team members, peers, and leaders to identify steps that can be taken to unleash the energy, intelligence, and creativity from within the entire system in pursuit of the shared reality they hope to create. The connected GAL practitioner has relationships that welcome the involvement of many and tap the intelligence of all. The connected GAL leader is an active listener, good communicator, and a great collaborator, appreciating the diversity of ideas and opinions that exist within the system.

The courageous GAL advocate tackles wicked-hard problems with team members, peers, and leaders, demonstrating the tenacity to take risks and the guts to be decisive, often in the face of great opposition. The complete GAL practitioner is compassionate, patient, and kind with all partners to create an energy between people so they feel seen, heard, and valued. The GAL practitioner is open to doing the requisite personal work toward self-realization to grow toward a place of wholeness and has an abundance mindset to help others to develop their highest potential.

It is helpful to measure one's progress in these GAL values. The assessment I propose measures the GAL growth over two periods. The assessment captures the individual leader's perspective of themself combined with a 360 view from peers, team members, leaders, and partners using the seven GAL values and attributes. Figure 14 is an illustration of an individual GAL practitioner.

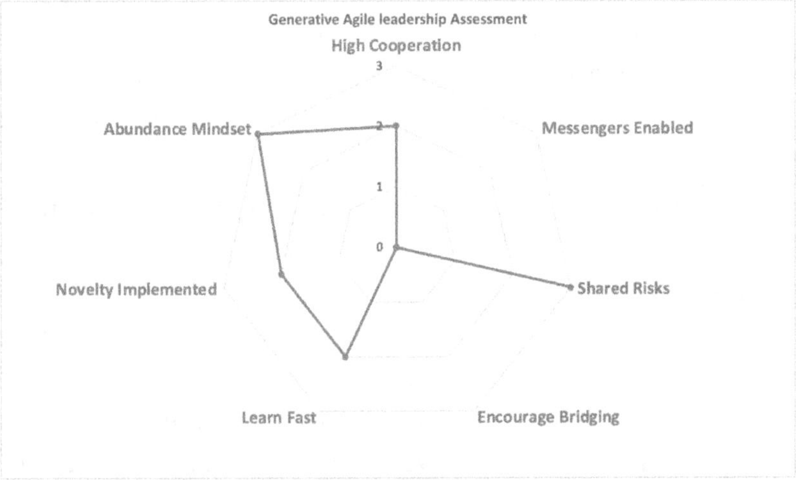

***Figure 14**: Generative Agile Leadership Assessment*

The assessment evaluates the GAL practitioner using the seven GAL values and attributes, each measured with a Likert scale of one (strongly disagree) to five (strongly agree). Zero represents a neutral position. The appropriate assessment times are when a commitment to the GAL practice begins and 90 to 120 days in the future to assess if the individual moved the needle by practicing the GAL values in the organization.

Leaderless Leaders

People can suffer economically, emotionally, and spiritually under inept leadership. Leaderless leadership is seen in the absence of leadership demonstrated by one or more individuals who are responsible for creating an environment for people and the organization to deliver value for customers and the organization's stakeholders and shareholders. The leaderless leader creates a context that is not beneficial for team members, peers, leaders, and partners. In fact, inept leaders diminish and distort their own self-actualizing reality. This leadership style is contrary to that of GAL practitioners and advocates that enable teams to be performance-based, resilient, and continually learning. The leaderless leader abdicates the responsibility of harmonizing the team or organization to become a thriving business.

It is estimated that disengaged people cost organizations in the U.S. more than $550 billion each year (Solis, 2015). Much of this loss is likely due to organizations allowing leaderless leaders to steal the innovative energy of people. Leaderless leaders fail to create an environment for people to deliver value for customers. My recommendation for leaderless leaders is to embrace and demonstrate the three traits of empathy, accountability, and delivering value to help enable happy contributing people to reduce the $550 billion loss.

Empathy

First, start with empathy for people. Walking in another person's shoes and feeling their pain is a powerful place for you to imagine their reality. Empathy is a staple in design thinking that helps leaders connect with other people's pain points and gain points. You begin by discovering the real pain of others by asking questions and showing up to have conversations. The discovered challenges can be turned into opportunities to provide an outcome of value.

Accountability

Second, be accountable and hold others accountable. Delegating is a wonderful art, but create a feedback loop that answers: how will we know when we are done? Set a timebox to have frequent conversations about progress. Share the results with others so they can learn from the experiences.

Delivering Value

Third, deliver value to customers and the organization's stakeholders and shareholders. My definition of value is a measurable outcome that can be realized and shared. When we understand people's pain and gain points, we can begin to describe and define the value to be delivered. Asking questions and receiving feedback will help to provide a more accurate value outcome.

The metaphor of gumbo provides a lens to see how leaderless leaders can find their mojo. Gumbo is a Cajun soup popular in New Orleans, Louisiana, USA. There are different types of gumbo to satisfy most consumers' palettes. The gumbo dishes include chicken and sausage; seafood; and vegetarian. My preference would be the vegetarian gumbo. The importance is to acknowledge the gumbo dish represents diversity and the need to make sure empathy and compassion are offered to everyone. In my research about gumbo and leadership, I discovered Morial (2020), who stated, "The strength and compassion New Orleanians needed in their many hours of need during the Katrina hurricane were missing in action from the leaders who could have made a difference." The discovery was that leaderless leaders were not acting to help all people during the Katrina hurricane crisis.

Leaderless leaders can affect the coalition of happy contributing people by missing the opportunity to serve the needs of the people who contribute to satisfied customers and thriving business.

In summary, the ingredients for the leaderless leader's gumbo include empathy, accountability, and delivering value. The practices of empathy, accountability, and delivering value are core to leaders but are not an exhaustive list of traits that leaderless leaders can use to be more effective in creating an environment for people and the organization to deliver value for customers and the organization's stakeholders and shareholders. The GAL values and attributes position the leaderless leader to build an organization that enables a thriving business.

LEADERS WANT BUSINESS AGILITY, TO MAINTAIN AND INCREASE GROSS MARGIN, AND TO LEVERAGE RISKS FOR EXPONENTIAL GAINS

The idea of practicing agility throughout the entire organization is a new phenomenon and is gaining traction in organizations of varying sizes. Business agility is the term used to describe the ability to optimize and align organizational missions across multiple teams of people performing important functions to enable the business to thrive. The business agility practice is a mindset, a set of values, guiding principles, and frameworks to deliver small increments of value for faster customer feedback.

The gross margin financial principle helps with understanding and monitoring product pricing, sales margins, and a healthy, sustainable business. The higher the gross margin percentage, the more money the business retains on each dollar of sales to service other costs and obligations. The ability to calculate a break-even point and evaluate the efficiency of producing the product or service enables the organization to set profitable prices that create a competitive advantage.

Starting and running a business is risky business. No pun intended! Starting from nothing with only a hypothesis requires the ability to leverage the known and unknown risks to pursue exponential gains. Business leaders want exponential gains to maximize investments and growth. I have not met a business leader who is satisfied with just breaking even. The startup business ecosystem is built on exponential growth with the understanding that at a moment's notice everything can become dust in the wind.

Unfavorable risks are waste (Japanese word "muda") that may cause the organization to drift off course and away from intended strategic goals. Organizations may intentionally elect to pursue risky opportunities to maximize profitability that can be beneficial to the bottom line. A delicate balance is needed to limit the potential exposure of employees, shareholders, and customers.

Diana Larsen, author of *Agile Retrospective* and co-founder of Agile Fluency Model

Leaders have a different awareness of their expectations for software teams and oftentimes a better vocabulary for talking to teams about what they're doing. Intact teams really dive in around, "What are we doing to build our capability and increase our capacity?" So, they're focusing more on that area. Coaches often have the experience of trying to do too much too soon and learning that they need to be more moderate in how they, in the game, the investments that they're making. That they need to focus on the flow of those investments more than trying to do everything all at once.

Dave West, former CEO, Scrum.org

I think ultimately the shift from the Industrial/Mass Production age to the digital age changes how we think about value. Mass production was, as the name implies, keeping complex work to a minimum and ensuring that a single blueprint is translated into 10,000s of products with low risk and in the most efficient way possible. In the twenty-first century, value is not the same. Yes, we still mass produce, but the value of a product or service is not in the mass-produced community bit; it is in the way that product is used by the customer and the relationship between the company that provides that product and the customer. That means the majority of work will be driven by creators, not managers of a zero defect/zero risk production system.

THRIVING BUSINESS FICTIONAL SHORT STORY

"This has been a remarkable turnaround," said Kathy, the CEO of Avant-Garde Realities (AGR). Her words rang out loud and proud through the spacious room where people lined up in folding chairs, eagerly awaiting her next words. "Our customer sentiment improved from being the dreaded meh to 70% of our customers shifting to being promoters and fans of our products."

Kathy paused and turned to Ashanti, the leader of these amazing developments. The young woman had worked hard since she had reached her position and was now proud to be recognized for it.

"Ashanti, congratulations on your one-year anniversary." Kathy beamed. "Your contributions to our organization have helped us become more agile and have led to us adopting this wonderful customer-obsessed culture. All of your work is greatly appreciated.

"To clarify, when I say agile, I am speaking about our teams' ability to deliver value to our customers and respond to market opportunities and threats. I'm so grateful to be able to celebrate the successes of the past twelve months. In the future, we'd like to see even more of the achievements we've experienced over the last year throughout the organization.

"As one of the founders of AGR, I still remember how this company started with a simple idea at a coffee shop. And now we have risen to be the best at what we do. There is no doubt in my mind that we can make AGR on par with some of the most successful companies that are now the talk of Wall Street and global financial markets. And it's all because of you."

The town hall attendees gave a big, roaring cheer with smiling, excited emojis flying on the video wall from virtual attendees. This type of hybrid meeting had become the new normal for organizations since employees were sent home to work. And by now, everyone was comfortable with this new way of doing things.

"Our business must be able to navigate the sea of changes in the industry and ensure continuity for the existing customer base," Kathy continued once the applause had died down. "Since Ashanti has done such a bang-up job with improving our customer sentiment and reducing the motion sickness issues in the fantasy vacation VR product, we will reward her with more work."

Chuckles came from the front of the room as other executives mumbled, "There is no rest for the successful." Ashanti looked proud of this, though. She was grateful to have the chance to show off the skills she had worked so hard to hone.

"The new entrepreneur and innovation initiative will now fall under Ashanti's guidance," Kathy continued. "This will ensure that we will be able to respond favorably to new opportunities in the market." Kathy turned back to Ashanti. "Would you like to share a few words about your new adventure?"

Ashanti stood up and took the mic, appearing proud and confident. "Wow, lucky me!" She winked. The attendees laughed at Ashanti's playful remark. "Truly, though, I'm such a lucky person to be rewarded with doing something that I have a great deal of passion for.

"I am a big community person, and our entrepreneur and innovation initiative will be extended to underestimated, local communities at first. People working in this new organization will be required to get out of the building and connect with the community in a hands-on approach. We will not only serve our interest but that of the community as well.

"You may be surprised at how many talented people are out there and may be great contributors to our future success. Our colleges and universities will also be fertile ground for new talent to ensure we have new thinking for our products. Stay tuned. Amazing things will follow in the near future."

Ashanti gave the mic back to Kathy, paused for a moment, and smiled while shaking her head as she went back to her seat, still not quite able to believe the success she had found in this company.

After the town hall meeting, Tarasi, Manny, and Marisela were assigned the leadership roles to review the internal and community innovation submissions. A flurry of new innovative ideas were submitted from AGR employees. Many ideas were focused on improving or extending the existing VR fantasy vacation product.

"We need scalable business ideas that move the needle," Tarasi said as she sat at her desk. She looked at a photo of her children, a reminder of who she was working so hard for. "Should we host workshops to teach about the concept of startups using the lean business canvas?"

"I went through several workshops," Manny answered, sitting across from Tarasi. "I would be willing to facilitate the learning experiences. Educating people about the great opportunity to launch a funded startup business as a leader is something that we must do a better job with."

"We can actually be the CEO of a subsidiary of AGR!" Marisela interrupted. "That would be so amazing! We can engage the organizational change management group to help us tell a better story, and we could create a campaign to raise awareness of our initiative, increase desire for this new opportunity, build up knowledge for startup business opportunities, develop the ability to create something from nothing, and reinforce the mindset for innovation."

"That sounds like an incredible idea!" Tarasi said, her mind flooding with all the wonderful possibilities.

Manny and Tarasi partnered to create an innovative idea called open metaverse. They were joined by Zen, a member of the local community with a passion for Afrofuturist visions. Afrofuturism allows people of African descent to tell stories of life in the future, thriving with joy. The movie *Black Panther* is an example of Afrofuturism.

The idea created by Tarasi, Manny, and Zen centered on the ability to bring your existing avatar to the metaverse community; use bitcoin, cash, or non-fungible tokens (NFT); leverage blockchain for immutable identity management; use artificial intelligence (AI); and support both virtual reality (VR) and augmented reality (AR).

The revenue model included a freemium version to build the virtual community, monthly subscription fees, and a pay-what-you-can option. The goal was to have an open metaverse for people to come and go with ownership of their data versus having their data stuck on one of the technology giant's platforms.

"We should call it Uwazi," Zen said, as they all sat in a conference room on a Tuesday afternoon. A wall of windows let them look down at the city buzzing with innovation. "'Uwazi' means transparent or openness in Swahili."

"I love that!" the team agreed.

"Now, how do we describe the metaverse in a simple language that people understand?" Zen pondered. "At first, I thought it was a virtual place for gamers to hang out, but it's so much more than that. To some people, the metaverse is a gaming platform, but to others, it's a virtual retail destination, a training tool, an advertising channel, a digital classroom, or a new gateway to digital experiences. The metaverse is a dreamer's paradise and open to whatever people's imaginations dream it to be.

"I guess we can say that the definition is still fluid. Now we have one of the largest social media platforms declaring a shift in strategy to adapt the metaverse as the next frontier for human engagement, which presents a whole new host of things to consider."

There was silence in the room for a moment as they all considered the meaning of this. Bright light splashed in on them, awakening their minds to all the possibilities.

"One thing we need to keep in mind is that we don't want the metaverse owned by just a few," Zen said. "We do not want the current state of the internet (Web 2.0) where only a few technology giants own everything, including our privacy."

"I think we should add Marisela to our team so that we bring her AI and machine learning skills to help us gain insights about the stages that customers take before becoming loyal fans," Manny said. "We should use the lean business blueprint to validate our hypothesis for Uwazi. We only need three areas of the canvas." Manny paused to pull a list up on the projector that read:

1. Offer (what product or service will you make and deliver to solve a specific customer problem?)
2. Customers (who are we serving?)
3. Value proposition (what customer need are you fulfilling? What is the unique promise your offer provides?)

"We can use this information to have conversations with potential customers to get their feedback. This would provide us with some data for customer validation. Many innovators start with building a solution that is seeking a problem to solve. The result is wasted time and money because the customer's unmet needs were not discovered."

Marisela walked over and tapped Manny on his shoulder and looked at him with surprise. Manny smiled back at her and said, "We were just talking about you joining our team."

"I didn't realize that you knew so much about the process to start a business before you started a business," Marisela said in admiration.

"I learned this from a few lean startup workshops that I attended last year," Manny explained proudly. "Customers travel through several stages before they become fans. Customers can start as innovators, early adopters, early majority, late majority, or finally the laggards. That's why it is important that we perform customer validation to see if we can make money from the Uwazi innovation."

The team of four decided to divide and conquer. Tarasi and Manny were going to do customer validation with college students at the local university while Zen and Marisela would hang out in Zen's neighborhood and hit some coffee shops and diners. The goal was to interview 100 people. Each team would have conversations with fifty people to obtain feedback about the Uwazi innovation idea.

Tarasi and Manny headed down to the university and used a lemonade stand to attract students for a conversation. The lemonade was free, and they were able to speak with more than 100 students. The unmet needs shared by the college students included:

1. Better and cheaper games
2. The desire to own their own avatar
3. A study room that is fun where they can share ideas
4. Create their own NFT for their art
5. Some shared utopian ideals of virtual world peace
6. A better immersive learning experience
7. The ability to travel to other places

Zen and Marisela's journey was a bit more challenging as they ventured to a local coffee shop called Hangout. The smell of roasting coffee paired with soft music and low lighting seemed perfect for creating the ambiance needed for conversation. But the patrons did not want to stop and answer their questions.

Marisela threw her hands in the air and blurted out, "You know what the problem is? These people are unable to visualize the opportunities or needs. We can use our design thinking skills to create a few low-fidelity prototypes that we can test and get feedback on instead."

"I'm up for any ideas that will get people to talk to us," sighed Zen. "A picture is worth a thousand words. We can draw a few pictures with a headline to tell the story of each image."

Zen drew a picture of Snoop Dog's NFT called Snoop Doggie coins in the metaverse with the headline "Snoop Makes Millions." Marisela followed with a school, a world, and the headline "Learning Anywhere With Anyone." The pair created a total of seven images and found a spot in a cozy corner of Hangout so they could have conversations about unmet needs.

Sherl, a local artist who frequented Hangout, saw Zen and walked over to the corner, eyeing the posters critically.

"Are you trying to be a mime with bad art?" Sherl laughed after introducing herself.

"It's our best effort." Zen laughed. "Admittedly not great, but it starts conversations, and we want to listen to everyone's input. As an artist, what are some of the biggest pains that you experience?"

"Zen, you know the struggles," Sherl said. "It comes down to selling your art, man."

Marisela pushed her way between Zen and Sherl and offered up a big smile. She introduced herself, then asked, "Do you happen to have any children? We'd really love to get feedback from parents and guardians."

"I don't have kids," Sherl replied. "Though I often babysit my sister's kids. Maybe I can still be helpful."

"You sure can," Marisela said. "Anyone who spends a decent amount of time with children can help us on this. Would you mind looking at my picture 'Learning Anywhere With Anyone' and tell me what unmet needs this idea could help with?"

"Sure," Sherl shrugged. "I can see how a secure, immersive space for learning could help while I'm babysitting my nieces and nephews by teaching them something useful besides watching animated cartoons."

The conversation continued for a little longer before Sherl had to get going. They thanked her for stopping by, then asked her to send some friends over since she knew so many people.

"I can do that," Sherl agreed. "I appreciate the time and effort you're putting into getting to know your customers better. So, I'll spread the word."

It took Zen and Marisela several weeks to get the fifty customer validation interviews. The good news was that they were able to visit various stores at different times to find a wider range of people for feedback.

The Uwazi innovation was accepted as a startup business venture and received a pledge of $1.5 million to build a prototype to test in the market. Ashanti created an advisory team to guide Tarasi, Marisela, Manny, and Zen as they assumed C-level roles in the new venture.

The open metaverse concept was launched seven months later, and within a year, it quickly grew to one million subscribers. Half were freemium subscribers who used the free tier as they explored if this was a fit for their unmet needs. The remaining 50 percent were paid subscribers, dedicated to the concept of the open metaverse.

Tarasi, Marisela, Manny, and Zen presented the progress to the AGR C-suite, highlighting the customer traction and positive revenue the Uwazi initiative achieved. Harry, the CFO, asked if they were ready to launch as a business entity contributing revenue to AGR.

"Without a doubt," Manny responded. "We can't wait to see if this entity can be successful enough to spin off and become a force in the open metaverse."

Kathy, the CEO, stood up and started the cheers and standing ovation in celebration of the success. "Ashanti," Kathy said, "great job with launching our innovation program with such success. The results exceeded our expectations."

THRIVING BUSINESS - INFLUENTIAL LEADERS' VOICES

Diana Larsen, author of *Agile Retrospective* and co-founder of Agile Fluency Model

One good thing that agile does is it promises faster time to market, which can allow an organization to thrive against competitors. So, when you think of the fluency model and then all the different zones, I think of how this positions the organization to experience or even realize growth. Because if we think faster to market, there's an opportunity to acquire new customers or even retain some customers that we have. I'm just looking to see how the fluency model helps to shape that a little bit or influence it in some way.

In some organizations, the way they have ... I mean, the model is a team model. So, we're thinking about basically individual team fluency as we're going along. Of course, as organizations get larger, they have many teams and they're doing different things. I've talked to several folks who have said, "In our organization, we have focusing teams, and they work mostly with our marketing department or on short-term initiatives that we need to get out to the public awareness," or whatever that might be. And they do quick turnaround kinds of things.

Then we have delivering teams—living zone teams, I guess—that work on our flagship. Continually improving our flagship product as our customers tell us what the next thing they would like to get is and how quickly they need it, and those kinds of things. Then, those organizations will often have some optimizing zone teams that are doing their R&D, that are out there testing the boundaries, that are looking for new product opportunities or new customer areas or how can we use this product to serve those folks in a different kind of way. Those are tied teams that need all the different skill sets in them to be able to ideate in a good way. Maybe they're using ... design thinking kinds of practices to help them do that, continually creating those very, very tight feedback loops for testing their ideas.

There are several organizations that we've worked with that will have teams ... will identify that they need teams in more than one cell. That helps them understand what investments they need to make. So that works well. Rather than thinking ... In the early days, we used to get people asking us, "Is this a delivering organization?" We always had to correct them and say, "No, no, no. This is not a model of organizations." It may have some resonance with other organizational models you've heard of, but this is about team behaviors. Organizations may need a variety of team behaviors to get their work done, to serve their customers.

This is excellent. Walking through the context of values to happy contributing people, satisfied customers, and thriving business. I think this gives us a good context of how the model can be used from a team perspective and how those teams could influence change and add value to the organization. I really, really appreciate the context that you've provided for us.

Howard Sublett, former CEO, Scrum Alliance

A thriving business. What it's not is revenue generation—that cannot be the number-one marker of a thriving business. Revenue generation and profitability should be the byproducts of a thriving business. Businesses can be platforms for change in the world. A thriving business knows why they exist and who they're there to serve. They understand what the product or service that they have can be and how it can be impactful to the world. With a mission focus and a clear why statement for the business, employees are happier and engaged. Customers want to engage with that kind of company. The byproduct would be a healthy revenue generation. Too many businesses get caught up in profit and loss statements. P&L statements are lagging indicators of a just cause.

We live in a world where whatever we're going to make or manufacture needs to have a larger purpose behind it. People are making lightbulbs that last fifty years and consume less electricity. They aren't making them because people need light; they are making them because these lightbulbs can help save our environment—it will make a difference in the world. Those are the kinds of businesses that will not just survive but thrive in this economy.

Marty Nelson, Alchemy Code Lab founder

A thriving business has market momentum, a good work environment, and uses its resources wisely. I have noticed that profitability and revenue have an outsized effect on the more human-centric things we value. What we really want are those things like meaning and belonging and good work relationships, a good "culture," but without the sustenance of cash, things go sour pretty quick. But we need that good culture to do the kind of work that produces high value. So, thriving has to do with being in a virtuous cycle.

I was working with one company doing a lot of merger and acquisition to transform a set of point-solutions into a cohesive suite offering. It was a great vision and story. But I think top management was focused on the exit, on the buyer who would come in down the road. So, the company was good, but not great because the vision was a means to reach a different end. Things worked out, but I never felt they thrived at the level they could have.

Dave West, CEO, Scrum.org

To me, a thriving business describes that innate energy created between the three elements of customers, team, and market. When your customers love what you are doing. When the team is working really well together. And when the market respects you. To create that energy requires a great work environment, a clear mission, a product that can be shaped, and lots of luck! It also comes and goes within a business because those elements are outside of your control and non-deterministic. Customers might love what you do one minute and then not the next because their needs change or the people you are working with have other issues. The work environment can be undermined by bad people or a lack of attention by leadership. The market is fickle and can suddenly change as everyone focuses on a new darling. Of course, if I was Steven Blank, I would say, "All of those things are to be expected; it is therefore important not that they happen but how you respond to them." Our ability to be agile is not from our ability to deliver, but our ability to react to the result/learning of our delivery.

I have been lucky to be part of three thriving organizations. The first one, Rational Software, where we had a clear mission, customers that loved us, and a team that cared. Even if, at times, we were kidding ourselves, we still had a higher mission of improving the industry. Our leadership did a great job to balance the financial needs of the company with our own goals. We had lots of formal and informal feedback loops. And, at the end of the day, we were proud to put on the shirt and say, "I work for Rational Software." The second one was Tasktop, a startup. A very different size from Rational with less than twenty employees when I joined. But we all felt a kinship, like Rational. We believed in the mission and for our fellow teammates. That translated into energy and passion, which the market respected. We also lived the values of being kind and humble (even if I failed a few times in that one). The third one is what I am part of at Scrum.org: the combination of a growing number of customers taking our training, passing our assessments, or reading materials on our website with a great community of PSTs and teams who are sharing ideas and supporting each other, and a market that respects us and what we are trying to do—even if they do not always agree with our approach.

At the end of the day, I was proud to work for all of these organizations. We were not perfect, but they combined mission with values, creating an environment that encourages a certain type of person to join, which re-enforces those values.

Agile is Emergent and Incomplete

Failure is feedback, so iterate and deliver small increments of value.
— Dr. Dave

The agile manifesto began by stating, "We are uncovering better ways of developing software by doing it and helping others do it." I would change "software" to "software, products, and services" to illustrate that agile can be used to develop products and services beyond software. Emergence does not mean that intentional and conscious planning is not applied. It means that there is space for discovery and the freedom to adapt after inspecting the deviation from the desired outcome. For example, a system architecture can be provided, but after discovery, changes can be made for improvement or to halt moving forward with a faulty hypothesis.

The agile mindset is one of a journey, and destinations are experienced through delivering incremental value to customers by empirical methods. The ability to engage in spontaneous collaboration that may produce an increment of value for customers and the organization is unparalleled. A fundamental effort is to make all of the work visible to the people responsible for delivering to ensure a common understanding. Failure and success are equally available to members of the organization. Agile as an emergent practice allows people within the organization to experience freedom that ultimately increases the ability to be more innovative.

The concept of agile being "incomplete" is a simple way of saying that agile is not a silver bullet and requires additional inputs, practices, and theories from other disciplines. The agile umbrella is a container for many frameworks and practices that help teams discover emergent ways to deliver value, but it is incomplete to satisfy the organization's full needs to experience business agility. Business agility is the vision and desire to apply agile practices and frameworks throughout the entire organization. The popular agile frameworks (that include Scrum, Kanban, and Scaled Agile Framework) have a primary focus on delivery. Design thinking and human centered design (HCD) are not part of agile practices and frameworks, but the outcomes from these practices provide tremendous clarity on customers and their behaviors with the organization's products and services. Armed with the outcome from HCD, teams are able to make decisions that will delight customers.

The incompleteness of agile is a good thing. Over the years, I have experienced the introduction of mindfulness into the mindset of agile practitioners, and people have benefited from the contributions. The application of empathy and awareness of the needs of others and oneself is incredibly freeing. We can now discuss how we feel versus what we think, which is open to both men and women. Agile leadership is another practice that emerged in recent years and is now available as a course and certification.

Agile Mindset

Agility is now mainstream. The goal for progressive organizations is to apply agility holistically throughout the organization. Aghina, Handscomb, Salo, & Thaker (2021) stated, "One of the measures of an organization's success is its agility—whether it manages to stay at least one step ahead of its market." The growing phenomenon of companies seeking to adapt to this new way of working is expanding daily. I spoke with Howard Sublett, Scrum Alliance's former CEO, and discovered that there are more than one million people globally with a Certified Scrum Master (CSM) designation. Scrum Alliance is one of the professional development organizations providing agile certifications in the world.

To practice agility, companies must do the following:

- Deliver frequent customer value
- Inspect and adapt—pivot for the maximum value
- Minimize waste (muda)
- Support collaboration and innovation
- Engage in continual learning
- Create psychological safe spaces for emergent leadership

The journey to agility begins with a mindset and not frameworks. Ray (2015) said, "The mind is a flexible mirror, adjust it to see a better world." The people making up the organization must first shift thinking to see defining a set of values as the primary goal to pursue. The agile manifesto begins with four values that speak to people's interactions, delivering working products, customer collaboration, and responding to change to deliver value to the organization and customers. The mindset shift is a culture shift that brings about a new set of norms and beliefs that people can rally around. In a survey, Miller, Okamoto, & Page (2016) discovered that "businesses are reinventing themselves to operate as networks of teams to keep pace with the challenges of a fluid, unpredictable world." Autonomy and self-organization are staples for teams within the organization to focus on mission, product, market, or integrated customer needs. The goal is to have small and nimble teams in the organization designed to partner in a network of teams. McChrystal (2015) described the relevance of shifting the approach to deploying small empowered teams to make decisions within the context of the bigger vision to achieve the desired outcome.

BUSINESS AGILITY (BA)

Business agility (BA) is shown in an optimized organization capable of responding to market dynamics to take advantage of opportunities and threats that enable business and individual growth. We use the lean principle of "optimize the whole" to indicate all value streams are optimized to maximize value, minimize waste, and increase flow. BA is a value delivery system that satisfies change caused by customer demands, competition, market dynamics, and regulatory requirements. The goal is to optimize the whole organization and not sub-optimize by doing only one business functional area. The optimize the whole concept is one of the seven principles of lean software development practices. The BA mindset must be culturally implemented at the organizational level and not just in the technology groups, specifically information technology (IT) software development. When agility is limited to the IT department only, the theory of constraints is activated in the rest of the organization. Goldblatt (1990) stated, "System constraints occur when anything limits a system from achieving higher performance versus its goal." The goal is to thrive and not just survive.

Organizations have several business functional areas that include marketing, sales, finance, and IT to carry out critical activities to ensure profitability and sustain the organization. The business functional areas must be transformed to become BA enabled to increase the effectiveness of the organization. These groups operate as silos and mostly engage in transactional relationships and interactions. How beneficial it would be to have true cross-functional teams of skills and knowledge contributing to a positive outcome of value for an organization initiative.

Happy contributing people are the catalyst that enables BA to grow sustainably. Conner (2010) stated, "Nimble organizations have a sustained ability to quickly and effectively respond to the demands of change while continually delivering high performance." The idea of becoming BA-capable is a great desire by many organizations. Startups and established businesses need BA to thrive in an uncertain and turbulent environment. The ability to satisfy customer demands is paramount to profitability, and the only way to achieve this goal is through agility, because customers demands are not static (Economist Intelligence Unit Limited, 2009).

Figure 15 illustrates the required domains when implementing BA in the organization.

***Figure 15**: Business Agility Domains **Adopted from:** Business Agility Institute (2021).*

The four BA domains are: customer, the heart of business agility and sustainable business; leadership, providing support for the business transformation; individuals, innovating higher quality products and services; and operations, sustaining digital systems availability. The first domain, customer, explores the relationships with the workforce, partners, and board. The second domain, leadership, emphasizes one team (one dream), people enablement, and strategic agility. The third domain, individuals, focuses on craft excellence, growth mindset, and ownership and accountability. The fourth domain, operations, zooms in on structural agility, process agility, and enterprise agility. These four domains can be implemented using a blend of design thinking and agile frameworks, and they can co-exist with traditional frameworks.

Business Agility Transformation at a Global Pet Parent Retailer Client

One of my clients is a global pet parent retailer with a transformation goal to adopt agile at scale to achieve outcomes that include delivering solutions faster to market with quality, aligning the organization's strategic objectives with the digital product development teams, and digitizing critical customer-facing solutions to reduce friction in the customer experience. As a business, executive, organizational, and agile coach and trainer, I have the desire to begin working with the leadership team to understand the challenges experienced in the organization, then transition to coaching the digital product delivery team(s). That is not the reality experienced because leaders see the challenges with the people doing the work, and the leaders do not think they are part of the challenges. It is human nature to not want to be labeled as a challenge and impediment to success.

The problem was the standard complaint that IT does not deliver digital solutions fast enough to respond to market shifts and customers' unmet needs. The organization wanted to pursue the promise of the SAFe benefits described as 30% happier, more motivated employees; 50% faster time to market; 35% increase in productivity; and 50% defect reduction (Scaled Agile, Inc, 2022).

The approach was to train everyone involved with the transformation, establish a lean-agile center of excellence team, create a roadmap for the transformation journey, define value streams, and develop an agile product management practice to move toward being more customer centric.

The outcome was the ability to pivot during COVID-19 in 2020 to use digital tools for pet parent curbside pickup and drop off, and it averted $800K in monthly revenue loss. In the United States, national television ads were shown of a young woman informing her mother that they were able to book an appointment using a mobile device, reducing the friction of calling to schedule an appointment for their pet. A shift in the organization behavior included the C-level team attending program increment (PI) planning events during the SAFe transformation journey.

Business Agility Activators/Influencers

BA is activated or influenced by external factors that impact business functions. To achieve the maximum outcomes, happy contributing people are at the center of delivering value through improved delivery flow and the application of BA enablers. Happy contributing people create the innovations that support nimble navigation of wicked-hard problems.

Customer demands, competition, market dynamics, and regulations are the primary activators or influencers for an organization to engage in change. The lean startup and lean business canvas models emphasize the need to understand customer demands to decide if the organization should stay the course or pivot with the product or service direction. Validating the product direction through awareness of customer demands would result in the appropriate measurable outcome. All other efforts are risky and should be done based on clarity of customer demands to achieve the desired goal. Innovation is based on human desirability, technical feasibility, and business profitability in the design thinking paradigm. The knowledge of customer demands in relationship to human desirability informs the organization of the decision path to follow.

Every organization experiences competition from comparable or startup companies. Competition not only exists from local companies but also from global competitors. Christensen (1997) defined disruptive innovation as startups that build less expensive products that displace the incumbent as the leader in a specific market. Every company must always be mindful of disruption that may occur from a startup or established company. The Netflix DVD-by-mail business case represents disruption to the now defunct incumbent Blockbuster DVD and VHS rental company.

Market dynamics are external factors beyond the organization's control. For example, an increase in the oil industry adversely affects the price of food. United States President Joe Biden, in response to high gas prices, released 1 million barrels of oil per day from the federal reserves for a temporary time period to dampen prices from increasing (White House, 2022). Additional pressures were placed on the oil and gas industries to reduce prices.

Regulations are specific to certain industries that include healthcare, financial organizations, manufacturing, and others.

The Food and Drug Administration (FDA) is a United States government agency responsible to keep food and drugs safe for consumers. Regulatory bodies often validate compliance to specific policies through an audit. ISO-9001 is one audit standard that specifies requirements for a quality management system. Organizations use the standard to demonstrate the ability to consistently provide products and services that meet customer and regulatory requirements.

Business Agility - Influential Leaders' Voices

Diana Larsen, author of Agile Retrospective and co-founder of Agile Fluency Model

Business agility is for me as Joshua Kerievsky lately has been talking about: quick, easy grace. For a business to ride the waves of customer desire, right, and customer service or serving our customers in the way they want, in a way that is resilient to all the changes in our economic environment, our marketplace environment, and you know, in a way that lets the business thrive and serve its employees and its community and its customers.

I'm a child of the '60s, and so there was a group of folks that I used to hang out with when I was a teenager and a young adult, and one of our kind of watch phrases was, "Well, you just have to roll with it." Harrison Owen, the creator of Open Space Technology, used to call it wave riding. You have to be able to roll with it. That really is the nature of our business marketplace now. Do we have the flexibility; the quick, easy grace; the versatility to roll with it and to roll with whatever comes our direction: changes in our customer, customer ideas, changes in the regulatory environment, whatever those may be? Can we respond and even anticipate those changes and continue to kind of roll with it, keep going, keep moving along?

The way the agile fluency model fits with that, I mean, the agile fluency model is a model about team skill proficiencies, team behaviors, and that's what we're describing in our model: these certain kinds of team behaviors that generate certain kinds of business benefits. It's a model with enormous organizational and management implications.

You can't just point at the teams and say, "Be different." We live in systems, and one of the things we know about systems is anytime we make changes in one part of the system, it is going to affect the other parts of the system in some ways, often unexpected, often emergent, over time. Being able to make the adjustments, to inspect and adapt at the business level, and understand what that means for what we need from our teams and then manage and organize ourselves to do business in a way that allows us to roll with those changes is, for me, what really is business agility.

Dave West, CEO, Scrum.org

My experience is that many executives want their organization to become more "agile" but do not have a clear definition of what it is. At best, they believe that they want their organization to deliver change to their customers (and new customers) faster. Time to market is their number one measure. But when pressed on what will change to support this increased time to market, they talk about process and maybe technology.

For the poster children of business agility (ING, Fidelity, John Deere, etc.), the reality is much more complex than adopting a new way of working, and the change they have achieved is not uniform with parts of their business being far more agile than others. All of these enterprises state that there is no single blueprint for becoming more agile, but instead some examples to learn from. These examples tend to be clustered into the following groups:

- Organizational models – How teams and teams of teams are aligned to deliver value and supported in that process.
- Management practices – How to encourage responsibility, support transparency, and enable creativity with a set of people who have not worked in this way.
- Planning and funding practices – How do you plan and budget for this way of working?
- Team practices – How do teams work, and what happens when you have several teams with large amounts of dependencies?
- Technology – What technologies enable scale without scaling?

From my experience, success tends to depend on two fundamentals. Clear alignment to the customer and focus. Alignment seems easy; after all, teams have to have clear objectives. But often, I see most teams not be able to describe the outcome they are seeking or the customer they are serving, to not be able to describe the customer in terms of why they are using the product or feature. Also, teams often are serving many customers, which is reflected in the quality of the things created and the lack of motivation and flexibility. Focus applies to both the way in which teams are aligned, but also what organizations are doing. Modern enterprises seem to have a hard time making decisions and say yes to everything. This results in a complex spider's web of priorities and dependencies. Leadership should make decisions and provide clear direction and focus. They should define priority and value.

Howard Sublett, former CEO, Scrum Alliance

The move to business agility is the awareness that agile frameworks and principles don't just exist in software development. For agility to thrive, we need to change what happens in the rest of the business, in HR and marketing for example. There's not a wall there.

In the early days of the agile movement, too many businesses thought that Scrum and other agile practices were just something the dev teams did; that it didn't affect them. But it does.

With that being said, business agility doesn't mean that every aspect of the company has to run in exactly the same number of sprints and releases. There are things in many companies that have long tails or that are booked out years in advance. That may feel anti-agile, but certain things have to be done that way for the trajectory of the company. So, the whole company doesn't have to run at the same cadence.

At Scrum Alliance, we've just reorganized in a way that embraces business agility to a level that maybe no one else has done or just a few businesses have done. We're stretching the idea that working in a Scrum rhythm involves the entire organization. We've moved it beyond the concept of cross-functional teams within departments—marketing teams or IT teams. Today, all of our teams are cross-functional across departments—maybe we should call them cross-departmental teams. In other words, all of the pieces—from marketing, education, IT, events, and so on—needed to deliver a product or service are present on each team.

And each cross-departmental team is radically focused on the needs of a particular customer segment. They're not just thinking about the value of a product; they're thinking about what value they can provide to a customer segment.

We've changed our leadership structure as well. In my prior role as the chief product owner, I was paired with my equal partner, Melissa Boggs, as chief ScrumMaster. We are one leader in two people. And together, we are radically focused on the overall product we're delivering as a company. I'm thinking more about what and why we deliver and she's more concerned with how, though she has a lot to say about the why as well.

To me, this new, evolving, emerging model may be the most idealistic, hopeful view of what business agility will look like in the future.

Mike Cohn, Mountain Goat Software

I think a business is thriving when it is providing its stakeholders with what they want. That would typically include providing customers with products that better meet their needs, providing owners with something like a monetary reward or a psychic reward for doing good in the world, and providing employees with positive experiences and pay.

Business Agility - Scaled Agile Framework (SAFe) Model

Business agility in large enterprises requires a multidimensional approach that includes customer centricity, lean-agile leadership, investment governance, organizational agility, continual learning culture, team and technical agility, agile product delivery, and enterprise solution agility. My experience with BA and SAFe is that the transformation begins without lean portfolio management and enterprise solution delivery. This does not mean these two key areas are not important, but it signals the lack of readiness to implement due to a lack of skilled individuals to lead the work and the organization's inability to turn the page to implement new changes in digitization investments and governance. Scaled Agile, Inc. (2022) found "to achieve Business Agility, the organization requires a significant degree of expertise across seven core competencies." Figure 16 illustrates the SAFe seven core competencies.

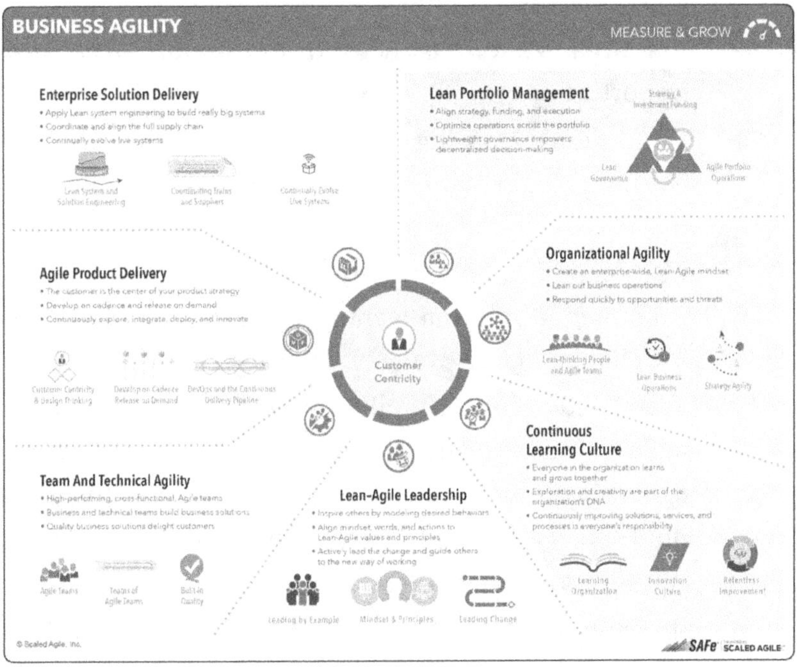

***Figure 16**: Business Agility **Adopted from**: Scaled Agile, Inc. (2022).*

My context is that there are eight core competencies with the addition of customer centricity. The customer is central to the existence of an organization, and without a customer-obsessed culture, sustainability and growth are limited. Developing a customer-obsessed culture and building an ecosystem for satisfied customers requires significant skills.

The eight SAFe business agility competencies are described as follows (Scaled Agile, Inc., 2022):

- Customer centricity - customer-obsessed culture that identifies the customers' unmet needs and fills them.
- Lean portfolio management - governance and authorization to invest in existing or new innovations.
- Organizational agility - the ability to quickly adapt the organization as needed to capitalize on new opportunities.
- Continuous learning culture - transforming into a learning organization by committing to relentless improvement and promoting a culture of innovation.
- Lean-agile Leadership - organization leaders being an example, adopting a lean-agile mindset, and leading the change to new ways of working.
- Team and technical agility - critical skills and lean-agile principles and practices that high-performing agile teams and teams of agile teams use to create high-quality solutions for their customers.
- Agile product delivery - a customer-centric approach to defining, building, and releasing a continuous flow of valuable products and services to customers and users.
- Enterprise solution delivery - describes how to apply lean-agile principles and practices to the specification, development, deployment, operation, and evolution of the world's largest and most sophisticated software applications, networks, and cyber-physical systems.

Business Agility Transformation at the Second Largest Automotive Manufacturer

The common problems that business operations people share about the partnership with IT is that the development cycle is too slow, there is a lack of visibility into the product development process, there are too many defects in software delivered in-house, and it costs too much for product development. This is part of the usual blame game, where one side of the organization lobs grenades at another group. Many factors contribute to the challenges of designing and building high-quality digital products. The reality is that IT digital products are one of the few assets in the organization that frequently provide scalable value. The other would be finance, which enables increase in revenue as money is invested.

One of my clients is the second largest automotive manufacturer in the world, and the goal of going on an agile transformation journey was to reap the benefits promised by using the SAFe framework. The SAFe benefits described are 30% happier, more motivated employees; 50% faster time to market; 35% increase in productivity; and 50% defect reduction (Scaled Agile, Inc, 2022). The specific challenges were IT was too slow, there was a lack of transparency and trust, and there was the usual blame game.

The approach to remedy the challenges was to create a coaching practice consisting of experienced agile coaches capable of coaching, training, mentoring, and facilitating the transformation for more than one hundred teams throughout the enterprise. It was not always that way. Before I arrived, it was rumored that saying the word agile as a way of working was a fireable offense. It took the steely nerves of several people to create the opportunity for the agile transformation journey to take place and continue to flourish. Training was provided in-house, where possible, to team leaders, delivery team members, human resources, marketing, the project management office, and other groups. Micro-learning courses were developed to allow for shorter learning experiences and immediate use in practice. An agile leadership cohort was developed to engage key leaders to learn what it meant to lead in a scaling agile environment. Furthermore, the content supporting the leadership agile mindset was implemented in key leadership courses delivered by the HR group throughout the enterprise.

The decision to create an IT transformation management office (IT TMO) to tackle emerging problems that require immediate focus using the SAFe practice was another benefit. The IT TMO direction was a collaboration between the senior leadership team, community of practices, center of excellence, and other areas of the organization in need of transformation. The emphasis was harmonizing ways of working (waterfall, adaptive, product management, etc.) that solved problems through experimentation and incremental deliverables. The outcomes were frameworks and reusable options for teams to opt-in at a sustainable pace. Many of the initiatives took on the inspect and adapt mindset by experimenting to deliver small incremental change versus a big bang solution.

The outcome was that the transformation was finally received and treated as an enterprise strategic goal by the CEO. Leaders began engaging in conversations about limiting the amount of work to increase focus on the important strategic goals. In the lean and agile world, we like to say limit work in progress. The transformation using SAFe, Scrum, Kanban, and other adaptive frameworks to shift to a lean and agile way of working has created increased transparency and trust between leaders and the delivery teams. It is not to say that everything is perfect and there are no bumps in the road, but the transparency enabled by psychological safety creates space for people to share failures without fear. It would be disingenuous to say that the agile transformation journey was the only cause for the shift. The great resignation and the limited amount of available people to fill key roles also was a contributor to functioning differently as well. This is a glimpse of the future of work: people working remotely and discovering ways to deliver value innovatively.

WHAT IS THE STRATEGIC PROBLEM?

The path to happy contributing people, satisfied customers, and thriving business begins with a vision. The vision is a response to a strategic problem that is either a threat or an opportunity. There is a double-sided sword that generative leaders must navigate to be mindful of business continuity for existing customers and innovating for future existence.

My travels have taken me to many organizations, and I find that the ability to describe the strategic problem is either non-existent or barely understood by the happy contributing people and far less by the satisfied customers.

The strategic process map provides seven steps to guide the approach when defining and evaluating one or more strategic themes. The journey of defining the enterprise strategy requires building a team to ideate and generate strategic possibilities. The team should be diverse so that many viewpoints are heard and applied. Figure 17 illustrates the steps used to define an enterprise strategy.

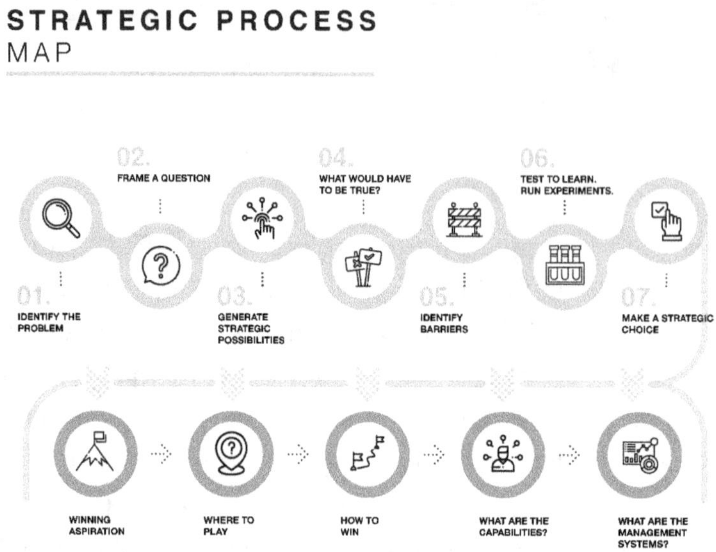

Figure 17*: IDEO Strategic Process Map.* ***Adopted from****: IDEO U (2019).*

Step one - Define the problem: Identify what isn't working. Describe what is blocking the organization from achieving the desired aspirations.

Step two - Frame a question for the problem: Turn the problem into a question to generate more possibilities. Start the problem statement with "how might we ..."

Step three - Generate strategic possibilities (be specific, concrete, and actionable): Collaborate with your team to identify as many ideas as possible that are specific, concrete, and actionable. Capturing responses to the question of "how might we ..." will get increased representation from a diverse community that could be facilitated through an innovation session to capture the diverse voices and ideas. A virtual tool could be used by the facilitator to support individuals remotely and in the same location to place their ideas on a board for conversation and review.

Step four - What would have to be true: Discover what would have to be true to realize the strategic possibility.

Step five - Identify barriers: Identify barriers that would make the strategic possibility unattainable or unsustainable. Some of the barriers may include people who are fearful that their privileged positions may decrease creating resistance to the change in being considered first for everything in the organization.

Step six - Test to learn, run experiments: Design, implement, and test iterative prototypes to have empirical results to increase knowledge and experience. Create small experiments that are ten to thirty days in duration to ask what is working and not working. The shorter feedback loops will help to verify your hypothesis about the validity of the experiments.

Step seven - Make a strategic choice: Make a choice between the strategic possibilities that will be selected for implementation. The strategic decision is based on the strategy choice cascade illustrated in Figure 18. The five strategy choice cascade outputs include: define the winning aspiration; define where to play; describe how to win; identify the existing capabilities that need to be undertaken to build a sustainable opportunity; and describe the infrastructure, systems, processes, and metrics that support and measure the strategy over time.

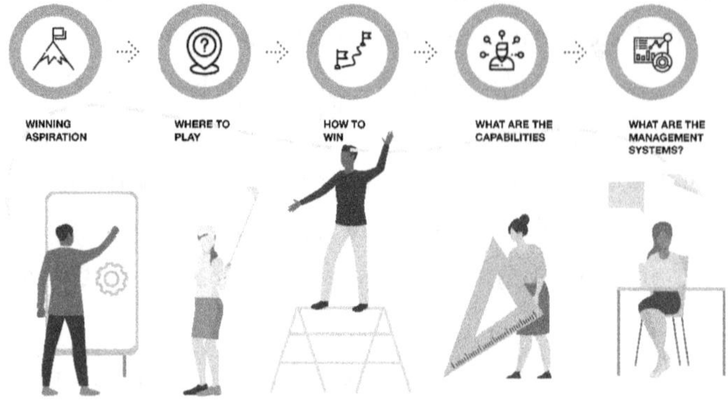

***Figure 18**: Strategy choice cascade.* ***Adopted from**: IDEO U. (2019).*

The strategy choice cascade serves as the output of your strategy work to visualize and share your strategy with others. The winning aspiration describes what to accomplish together and builds alignment with the organization's members. Where to play identifies the location, people, and product/service offering that potential customers will experience. How to win is the competitive advantage that describes the differentiators that will be realized by the strategy possibilities. The capabilities are the activities needed to build the competitive advantages to win. The management systems include technology platforms, organizational structures, training programs, and key measures.

IDENTIFYING VALUE WITH VALUE STREAMS

My first experience with value stream mapping was during my years doing six sigma black belt work at one of the largest banks in the world. People would confuse the lean practice for a martial arts activity. They would say, "Congratulations on receiving your black belt! What style of martial arts do you practice?" I had to be very clear that it was a lean manufacturing practice and not Bruce Lee Kung Fu. Obviously, they missed the words "six sigma" before "black belt."

The six sigma black belt practice used value stream mapping to identify an issue with efficiently delivering value to customers. A team of six sigma black belts, dressed in our corporate uniforms (tongue in cheek called karate gi) would invest time observing the way people worked and capture elapsed time for each step in the work process. During the observation, we would look for behaviors that may include people writing instructions or reminders on sticky notes to complete a specific task. Any steps that would reduce efficiency and create muda (Japanese word for waste) would be recorded. We would take up residence in the work area of the group of people for as long as four weeks. At the end of the observation, we would have a value stream map to review with the leadership team. Recommendations would be made for digital innovations as well as improving efficiencies on how people worked. Sometimes it would be physically designing a workspace for more effectiveness.

Value streams are becoming more popular for modern organizations moving toward an adaptive and product-centric operating model. Value streams provide a visual of the steps required to deliver value from the customer's initial request to the final delivery for use. The focus is on the customer and the time required to deliver value. There are two types of value streams. First is the operational value stream that focuses on the sequence of activities needed to deliver a product or service to a customer. Examples include manufacturing a product, fulfilling an e-commerce order, admitting and treating a patient, providing a loan, and delivering a professional service.

The second type of value stream is the development value stream, which focuses on the sequence of activities needed to convert a business hypothesis into a technology-enabled solution that delivers customer value. Examples include designing and developing a medical device, developing and deploying a CRM system, and launching an e-commerce web site.

Figure 19 illustrates an operational value stream for a loan process.

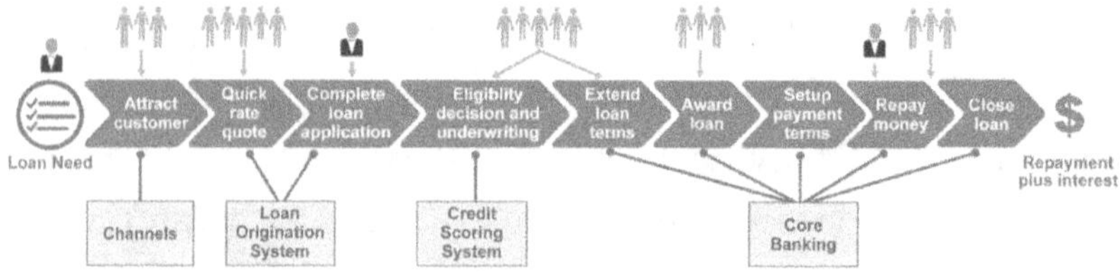

***Figure 19**: Value stream example of a loan process. **Adopted from**: Scaled Agile, Inc. (2022).*

What can be gleaned is the people who are involved in working on the steps of the value stream, the requesting customer's engagements, and key digital systems that help to automate the value stream. In a complete value stream, between each step in the value stream, a measure of the time elapsed is displayed that represents both muda (waste) and value. The purpose is to answer the "what, so what, and now what" questions: What is the waste that is observed and documented? So what, there is waste: what is the implication? Now what will we do to reduce the effect of the waste?

Measure and thrive is a mindset that organization leaders and happy contributing people must pursue to make radical decisions. Measuring the behavior of the value stream steps is essential to make decisions to stay the course or pivot for improved effectiveness. The measurement is learning about the customer-centric flow of value. Figure 20 uses the development value stream to illustrate the use of lead time, flow/cycle time, and flow efficiency.

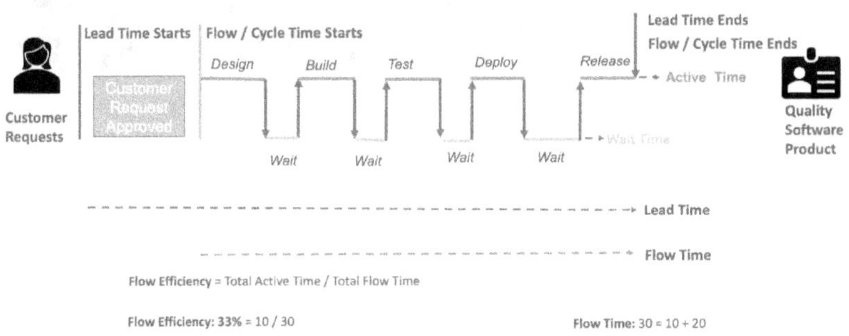

***Figure 20**: Development Value Stream with Flow Efficiency.*

Lead time is the elapsed time from when a customer feature request is approved to when it is released to the customer (flow state = done). A lag can exist from the approval of the customer feature request to when the development team commits to the flow item entering the value stream (flow state = active). An example is a customer feature request is approved by a lean portfolio group, and the development team then commits to doing the feature.

Flow/Cycle time is the time elapsed from when a flow item enters the value stream (flow state = active) to when it is released to the customer (flow state = done). An example would be the time it takes to deliver a new feature to a customer from when the feature is accepted. Flow/cycle time is a measure of how fast business value is delivered. Stewards of a business process or digital innovation process can answer the question of why things are taking so long.

Flow efficiency is the proportion of time that flow items are actively worked on compared to the total time elapsed. An example is when dependencies cause teams to wait on others, thus decreasing flow efficiency.

DESIGN THINKING AND EXPERIMENTS

Design thinking is an iterative process in which we seek to understand the user, challenge assumptions, and redefine problems in an attempt to identify alternative strategies and solutions that might not be instantly apparent with our initial level of understanding. Design thinking is a customer-centric mindset that employs empathy to best understand the customer's unmet needs. Design thinking is not limited to product design. Other applications include strategy design, business blueprint, digital transformation, and education. The goal is to provide business impact through design, fostering increased collaboration and innovation. Figure 21 illustrates the design thinking innovation flow.

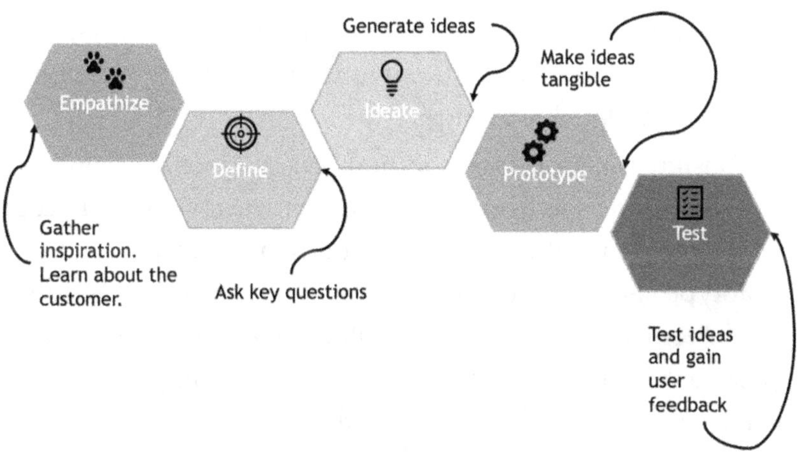

***Figure 21**: Design Thinking Flow* ***Adopted from**: IDEO U. (2019).*

The design thinking innovation flow is an iterative process that pursues short feedback cycles for faster learning.

The iterative process includes the following steps:

1. Begin with empathy to gather inspiration to learn about the customer.
2. Ask key questions of yourself and customers to help define the problem. This includes your users' needs, their problems, and your insights.
3. Generate as many ideas as possible by challenging assumptions and creating ideas for innovative solutions.
4. Make ideas tangible by creating prototypes. There are no limits to the prototypes, but a timebox is helpful.
5. Test ideas and gain user feedback.

The iterative process and short feedback cycles amplify faster learning about the customers' unmet needs and the possible innovations that are pain relievers and gain creators. Design sprints are a five-day iterative cycle of a team collaborating to empathize, define, ideate, prototype, and test a hypothesis. The end results would be multiple prototypes based on validated learnings.

Design thinking is a solution-based process. There is a problem space: we begin to discover a clear view and scope of the problem. There is a solution space: we develop and prototype several potential solutions until one or more satisfy the customer problem. You would move from the problem space to the solution space by doing the following: discovering insights into the problem, defining the area of focus, developing potential solutions, and delivering solutions that work. Figure 22 illustrates the transition from the problem to the solution space.

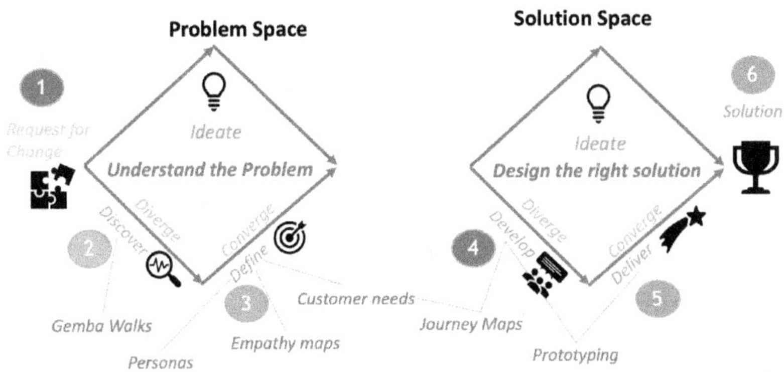

Figure 22: Design Thinking Problem and Solution Space. ***Adopted from:*** *IDEO U. (2019)*

The flow from the problem space to the solution space follows:

1. There is a request for change (customer or innovation)

Understand the problem

2. Discover the problem (ideas diverge): One activity is to go on a Gemba Walk to see the behaviors demonstrated by the people doing the work.
3. Define the problem (ideas converge): We can define a few personas, create an empathy map, and describe the needs for the solution.

Design the right solution

4. Develop journey maps and prototypes for evaluation (solution diverge)
5. Deliver a customer value (solution converge)
6. Implement the winning solution

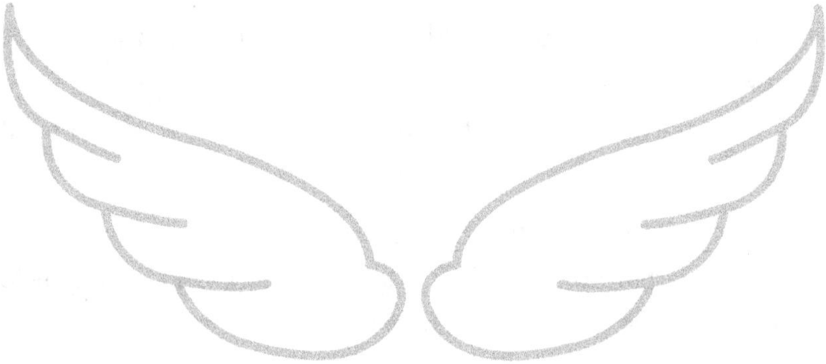

RESILIENT TEAMS

Resilient teams are the pinnacle of a thriving business. The journey to get within the sphere of a high-performing team starts with purpose (why) and continues to build toward sustaining the team's existence. The journey is going through the maturity realm of being low performers to high performers to resilient performers over time (Forsgren, Humble, and Kim, 2018). Occasionally, you may find a team that clicks immediately and becomes a resilient, high-performing team. Figure 23 illustrates the Grove model of team performance from creating to sustaining the team using a seven-step model.

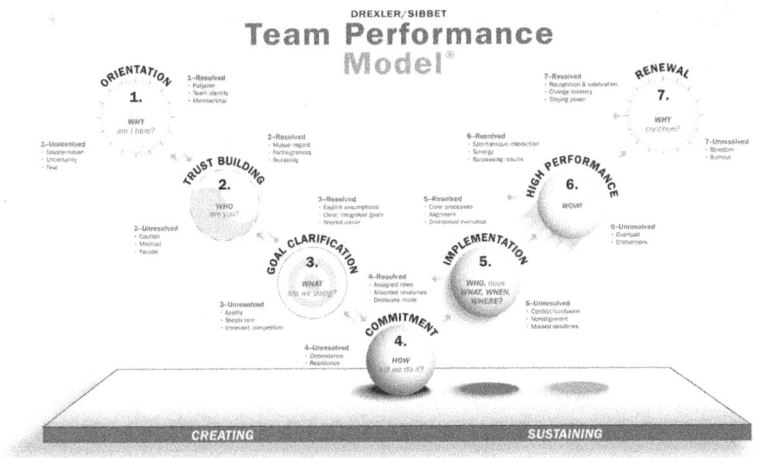

***Figure 23**: Team Performance Model. **Adopted from**: Sibbet & Drexler (2018).*

Steps one through four are associated with creating a resilient team. Steps five through seven contribute to sustaining the resilient team. Orienting the team members to the purpose of the team identity and the work to be done begins with initially stating why we are here and further discovering more purpose as time progresses. Trust building is essential for resilient teams. Covey (2008) said trust is a multiplier, accelerator, and energizer that is vital to reducing the tax associated with impediments that derail getting things done. The trust ecosystem includes self, peers, partners, direct reports, leaders, and community. Goal clarification for the team provides clear expectations and a shared vision. Assumptions are validated to ensure that the team is aware of the mission. With clarity, the resilient team can make commitments to delivering work incrementally. In addition, experimentation allows the team to learn how best to work. Implementation becomes easier with clarity and commitment because the resilient team is all in. Each member assumes a role and delivers to the best of their abilities. This leads to the wow factor of consistently performing at a high level. The trust level is high in the team and with leaders, partners, and customers. Finally, the resilient team identifies ways to experience renewal in commitment to the team by demonstrating staying power. When members decide to leave the team, it is in good faith.

SHIFTING CULTURE - NAVIGATING COMPETING VALUES

The thriving business must demonstrate the ability to be agile and innovative in the VUCA business environment. The organization must have the ability to focus on pivots that include flexibility versus stability and internal integration versus external differentiation to match the dynamics of internal and external factors. Internal factors may include shifting ways of working by aligning teams based on products over projects. External factors may include responding faster to change based on customers' unmet needs and market competition. The organization leaders must engage the team members to determine how to pivot between sustaining, transformational, fast, and incremental changes. Figure 24 illustrates the competing values framework for organizational culture shift.

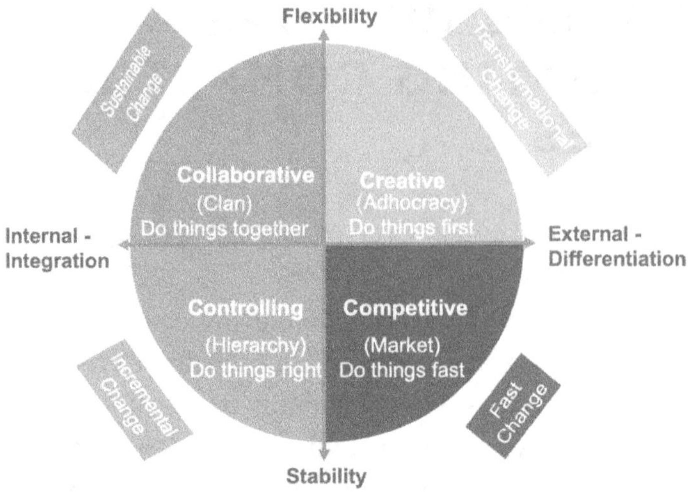

***Figure 24**: Competing Values Framework. **Adopted from**: Quinn and Cameron (2011).*

The concept of co-creating a culture shift with the organization leadership and its members may be a divergence from the status quo. However, it is becoming the new normal with the shifts in the newly forged relationship based on the great resignation and people empowerment in the workforce. The types of culture to consider may be adaptive and not static. The culture choices may be the adaptive integration of competitive (do things fast), collaborative (do things together), creative (do things first), and controlling (do things right). The opportunity to learn from patterns already explored by other companies can decrease the amount of time to mature during the culture shift. Examples include Amazon, competitive (do things fast); IDEO and Salesforce, collaborative (do things together); startup companies, creative (do things first); and pharmaceutical companies, controlling (do things right). Cultural intelligence capabilities will be essential to support the decisions of the organization leadership and members.

Modern organizations have a diverse business environment of happy contributing people with varying habits, beliefs, and assumptions. Cultural intelligence is demonstrating awareness of the distinct differences of the people working in the margins of the competitive, collaborative, creative, and controlling culture shifts that are associated with fast, sustaining, transformational, and incremental change respectively. Members of the organization should suspend judgment about the pivots, think clearly before acting, and run the experiment to learn (Early and Mosakowski, 2016). Navigating the cultural shifts is only made possible by applying cultural intelligence.

SUMMARY

A thriving business is dependent on happy contributing people, satisfied customers, and a clear sense of direction that aligns with the customer's unmet needs. This alignment positions the organization's leaders and teams to take advantage of opportunities by viewing the customer through an empathetic lens to connect with the pain experienced and gains desired.

Business agility is an important shift to apply the lean-agile mindset to respond to threats and also create innovations that delight customers' unmet needs. The context of business agility is optimizing the entire organization where appropriate, beyond the scope of IT departments. Imagine from the C-suite to the suite of teams delivering value faster to delight customers and increase market share.

Fundamentally, well-aligned organizations operate from a strategic problem perspective that is centered on meeting customers' unmet needs. Limiting the strategic work in progress (WIP) to between two and five items creates focus and limits the squandering of scarce resources and people's activities. It is why the question of "what is the strategic problem that needs our attention?" is important. Discovering the strategic problem and identifying options to pursue results is the best outcome. The outcomes are usually happy contributing people, satisfied customers, and a thriving business.

Design thinking is the exploratory approach to applying empathy to learn about the customer's unmet needs. The methods are simple and fun to use with a team of collaborative spirits. The goal is to create prototypes and test them, so that you validate learnings about each hypothesis's effectiveness.

Value streams make the business operations and IT streams of work visible, giving an opportunity to optimize areas to deliver value faster to customers.

ABOUT DR. DAVE CORNELIUS AKA "DR. DAVE"

Dr. Dave Cornelius, aka "Dr. Dave," is a business, executive, and organization coach, partnering with leaders and teams in for-profit, startup, and non-profit organizations to deliver a level of awesomeness that meets their goals. He works with Fortune 50 to Fortune 500 companies, helping them to deliver more than $1 billion in customer value. Dr. Dave is a coach to university researchers to facilitate the commercialization of their startup ideas that can be brought to market.

Dr. Dave delivers in-person and virtual immersive learning experiences in agile (Scrum, Kanban, SAFe, etc.), design thinking, lean startup, and professional speaking. His doctoral manuscript focused on "The Value of Scrum to Organizations" in support of his management and organizational leadership doctoral path. He also holds a BS in computer science and an MBA.

Dr. Dave is a big community contributor and the founder of the Agile for Humanity conference and meetup that supports the BIPOC community in having a voice in the product development and technology fields. He is the founder of the 5 Saturdays STEAM program that empowers under-estimated high school students to pursue knowledge and careers in STEM fields.

He is the author of five books *(Transforming Your Leadership Character: The Lean Thinking and Agility Way; Elastic Minds: What are you thinking?; Prayers to my Abba, Father, God; Belonging and Healing: Creating awesomeness for yourself and others; and Deliver Value: Happy Contributing People, Satisfied Customers, and Thriving Business).* You will find his monthly podcast and blog *KnolShare with Dr. Dave* on popular streaming services as well as KnolShare.org.

He is a husband, dad, and grandfather.

APPENDIX A: INTERVIEW PROFILES

Name:	Diana Larsen
Organization:	Agile Fluency Project LLC
What led you to your current position in your journey?	Early on in my career, I learned that work worth doing had to contribute to my personal mission. Once I had that in mind, I used it as my guiding principle to perceive the open doors and walk through them. So, everything I had done prior to fall 1997 had prepared me to accept the opportunity to attend an event that introduced me to agile, helped me to recognize it as an extension of my previous work, decide to choose it as my path, and find the ways I could contribute to the movement. More recently, I'm taking what I've learned throughout my journey and sharing it with others, so they can find more of their own new opportunities and open doors.
Who influenced your philosophy on life and business?	Many sources have influenced me: first and foremost, my grandmother, my mother, my aunts (lifelong teachers and learners all). As a young adult, my life was personally transformed by reading *Love Is Letting Go of Fear*, a book by G. Jampolsky. A university mentor, George Austin, and a consulting mentor, Chuck Ensign, introduced me to authors that helped to shape my business philosophy: M.P. Follett, E. Trist, F. Emery, R. Moss Kanter, E. Schein, D. Cooperrider, E. Deming, D. Hanna, and numerous other authors and colleagues contributed ideas and trainings in the early part of my career. For the past thirty years, I've been influenced by my collaboration with Sharon Buckmaster. I enjoyed newer influencers in the agile community, and in the recent decade, I've added new learning with G. Eoyang and D. Snowden about systems and complexity. All of these sources (and more) fed into my philosophy about life, work, and workplaces. I am also a product of the times I've lived in so far. And I'm looking forward to new influencers in the realm of virtual learning and internationalization/multiculturalism.

What would be your ideal form of work?	I'm doing my ideal work right now. Finding joy in learning, sharing, contributing, creating a more humane and sustainable world of work. Encountering interesting obstacles and finding the potential and possibilities in them.

Name:	Dave West
Organization:	Scrum.org
What led you to your current position in your journey?	It is a long journey. The short version goes something like this. I loved computers and programming and worked in dev at a large company. I saw a motivational speaker and found true north to 'help people deliver software just a little bit better.' I started with objects, went to Rational, then moved to RUP (met Ken). I moved to work with Ivar Jacobson (use case), wrote a book, moved to Forrester (analyst then practice leader), exercised my startup bug with Tasktop, and finally joined Ken at Scrum.org.
Who influenced your philosophy on life and business?	It started with Grady Booch and Jim Rumbaugh, then Ivar Jacobson, the idea of abstraction, the idea of systems and objects, and then the agile community, in particular Ken Schwaber.
What would be your ideal form of work?	Working with people I like on an interesting problem that has clear value to the world. Oh, and breakfast tacos every day! Hang on, that is what I do apart from the breakfast tacos!

Name:	Howard Sublett
Organization:	formerly Scrum Alliance
What led you to your current position in your journey?	You could say that everything led me to my [then-] current role as chief product owner for Scrum Alliance—but I believe it's more that this position came and found me. Getting introduced to agile for the first time by working at the Scrum Alliance all those years ago changed the trajectory of my career. Once I stepped in, I never wanted to step out. It became more of a passion than a job. And then, out of the blue, this position came and met me where I was. And when it did, I couldn't help but say yes, couldn't help but try. I believe in our mission; I believe in the Scrum framework. And I knew that if I didn't at least try to serve Scrum Alliance in this way, I'd spend the rest of my life wondering what if.
Who influenced your philosophy on life and business?	I'd have to say my father. He told me that I could do whatever I want, choose whatever career interested me, but that he expected two things from me: "Be a man of honor. Be a man of integrity. No matter what else you do, be those things." Being honorable and having integrity are critical pieces of my philosophy in both life and in business.
What would be your ideal form of work?	I love being in a group of like-minded people—people who are typically smarter than me and who share my passion to take risks, build great things, and advance a cause. Physically, I enjoy working in a collaborative space, with no walls and lots of whiteboards so that when things spark, anybody can get up and start drawing, writing, talking—building off each other's ideas. I detest fixed walls and rigidly scheduled meetings. A free-flow, open way to work is best for me.

Name:	Marty Nelson
Organization:	Alchemy Code Lab
What led you to your current position in your journey?	After two decades of software development and software development leadership, I happened to try teaching a night class for a code school, which led me to taking a full-time position. I was offered an opportunity to take over the code school in September of 2016.
Who influenced your philosophy on life and business?	Very influential have been people who led me to become a seeker, through different practices like yoga and shamanism. Also Anil Thakur, my boss when I was at Accelrys, who taught me development industry norms and demonstrated active respect and encouragement of all his staff.
What would be your ideal form of work?	The time and space to follow creative threads. It would feel like an adventure where there was a destination and goal, but very freeform exploration with time to craft and explore alternatives.

Name:	Mike Cohn
Organization:	Mountain Goat Software
What led you to your current position in your journey?	I got to the point I just couldn't work for anyone else. I'm entirely unemployable, so I started working for myself.
Who influenced your philosophy on life and business?	Probably everyone I've ever worked with.
What would be your ideal form of work?	Pretty much exactly what I'm doing right now or I wouldn't be doing it.

REFERENCES

Adams, R. L. (2016). "21 Famous failures who refused to give up."

Retrieved from https://www.huffpost.com

Aghina, W., Handscomb, C., Salo, O., and Thaker, S. (2021). "The impact of agility: How to shape your organization to compete."

Retrieved from https://www.mckinsey.com/business-functions/people-and-organizational-performance/our-insights/the-impact-of-agility-how-to-shape-your-organization-to-compete

Almquist, E., Senior, J. and Bloch, N. (2016). "The Elements of Value."

Retrieved from https://hbr.org/2016/09/the-elements-of-value

Bennett, N., and Lemoine, J, G. (2014). "What VUCA Really Means for You."

Retrieved from HBR.org

Barsoux, J., Wade, M., and Bouquet, C. (2022). "Identifying unmet needs in a digital age: A four-part framework for diversifying how and where to look." July-August 2022. Harvard Business Review. Retrieved from HBR.org

Briglia, C. (2020). "Why Customer Engagement Is Important."

https://www.gartner.com/en/digital-markets/insights/why-customer-engagement-is-important

Brown, B. (2015). *Rising Strong: How the Ability to Reset Transforms the Way We Live, Love, Parent, and Lead.* New York, NY: Penguin Random House, LLC.

Business Agility Institute (2021). "Domains of business agility." Retrieved from BusinessAgility.Institute.

Christensen, C. (1997). *The Innovator's Dilemma.* Boston, MA: Harvard Business Review Press.

Christensen, C. M., Hall, T., Dillon, K., and Duncan, D. S., (2016). "Know Your Customers' Jobs to Be Done." Retrieved from HBR.org.

Conner, D. (2010). "The Characteristics of Nimble Execution." Retrieved from http://www.connerpartners.com.

Covey, S. (2013). *The 7 habits of highly effective people: Powerful lessons in personal change.* New York, NY: Simon & Schuster, Inc.

Covey, S. M. R. (2008). *The speed of trust: The one thing that changes everything*. New York, NY: Free Press.

De Pree, M. (1989). *Leadership Is an Art*. New York, NY: Doubleday.

Deming, W. E. (1950). "PDSA Cycle." Retrieved from https://deming.org/

Dignan, A. (2019). *Brave New Work. Are you ready to reinvent your organization*? New York, NY: Penguin Random House, LLC.

Doerr, J., Page, L. (2018). *Measure What Matters: How Google, Bono, and the Gates Foundation Rock the World with OKRs.* New York, NY: Penguin Random House, LLC.

Duhigg, C. (2016) "What Google Learned From Its Quest to Build the Perfect Team." https://www.nytimes.com/2016/02/28/

magazine/what-google-learned-from-its-quest-to-build-the-perfect-team.html

Early, P. C., and Mosakowski, E. (2016). *Cultural intelligence.* Boston, MA: Harvard Business Review Press.

Economist Intelligence Unit Limited. (2009). *The Economist* article. "Organisational agility: How business can survive and thrive in turbulent times."

Retrieved from http://graphics.eiu.com.

Edmondson, A. C. (2019). *The Fearless Organization: Creating Psychological Safety in the Workplace for Learning, Innovation, and Growth.* Hoboken, NJ: Wiley & Sons.

Fleiter, T., and Plötz, P. (2013). "Encyclopedia of Energy, Natural Resource, and Environmental Economics." Retrieved from https://www.sciencedirect.com/.

Forsgren, N., Humble, J., Kim, G. (2018). *Accelerate: building and scaling high performing technology organizations.* Portland, OR: IT Revolution Press, LLC.

Greenleaf, R. K. (1977). *Servant leadership: A journey into the nature of legitimate power and greatness.* New York, NY: Paulist Press.

Goldbach, J. (2017). "Diversity toolkit: A guide to discussing identity, power, and privilege." https://msw.usc.edu/mswusc-blog/diversity-workshop-guide-to-discussing-identity-power-and-privilege/

Goldblatt, E. (1990). *Theory of Constraints.* Great Barrington, MA: The North River Press.

Granath, B. (2015). "Members of Apollo 13 Team Reflect on 'NASA's Finest Hour.'" https://www.nasa.gov/content/members-of-apollo-13-team-reflect-on-nasas-finest-hour

Gray, D., Brown, S., and Macanufo, J. (2010). *Gamestorming: A Playbook for Innovators, Rulebreakers, and Changemakers*. Sebastopol, CA: O'Reilly, Media, Inc.

Holmes, M. (2019). Microsoft CEO: "Every Company is Now a Software Company." Retrieved from https://www.satellitetoday.com/innovation/2019/02/26/microsoft-ceo-every-company-is-now-a-software-company/.

IDEO U. (2019). Designing strategy course.

Irwin, N. (2016). "What Was the Greatest Era for Innovation? A Brief Guided Tour."

Retrieved from https://www.nytimes.com/2016/05/15/upshot/what-was-the-greatest-era-for-american-innovation-a-brief-guided-tour.html?smid=url-share

Mäkinen, V., Adam, C., Ko, K. W. (2022). "The IT Operating Model Framework of the Future." Information Services Group, Inc. ISG White Paper.

Maslow, A. (2013). *A Theory of Human Motivation*. Mansfield Centre, CT: Martino Publishing.

McChrystal, S. (2015). *Team of Teams: New rules of engagement for a complex world*. New York, NY: Penguin Random House, LLC.

McGregor, D. (2006). *The human side of enterprise, annotated edition*. Englewood, CO: McGraw-Hill Companies, Inc.

Miller, D., Okamoto, T., and Page, T. (2016). "Organizational design: The rise of teams." Retrieved from https://www2.deloitte.com/us/en/insights/focus/human-capital-trends/2016/organizational-models-network-of-teams.html

Moore, G. (1991; 2006). *Crossing the Chasm: Marketing and Selling High-Tech Products to Mainstream Customers*. Harper Business.

Morial, M. (2020). *The Gumbo Coalition: 10 leadership lessons that help you inspire, unite, and achieve*. New York, NY: Harper Collins, LLC.

Osterwalder, A., Pigneur, Y., Bernarda, G., Smith, A. (2014) *Value Proposition Design: How to Create Products and Services Customers Want*. John Wiley & Sons.

Oswald, A.J., Proto, E., and Sgroi, D. (2015). "Happiness and Productivity." Journal of labor economics 33.4 (2015): 789-822.

Pink, D. H. (2009). *Drive: The amazing truth about what drives us*. New York, NY: Penguin House Group, Inc.

Prosci. (2016). "Organizational Agility as a Strategic Imperative." Retrieved from prosci.com.

Poppendieck, M. & Poppendieck, T. (2007). *Implementing lean software development: From concept to cash*. Boston, MA: Pearson Education, Inc.

Quinn, R. E., and Cameron, K. S. (2011). *Diagnosing and changing organizational culture: based on the competing values framework*. San Francisco, CA: Jossey-bass A Wiley Imprint.

Ray, A. (2015). *Mindfulness: Living in the moment Living in the breath*. Inner Light Publishers.

Reinertsen, D. G. (2009). *The Principles of Product Development Flow: Second Generation Lean Product Development*. Celeritas Publishing.

Schell, C. (2017). "Harnessing the power of diversity and inclusion for innovation." Retrieved from CIO.com.

Schmidt, E., Rosenburg, J., and Eagle, A. (2019). *Trillion dollar coach: The leadership playbook of Silicon's Valley Bill Campbell*. New York, NY: HarperCollins Publishers.

Senge, P. M. (1990). "The Leader's New Work: Building Learning Organizations." Retrieved from: https://sloanreview.mit.edu/article/the-leaders-new-work-building-learning-organizations/

Scaled Agile, Inc. (2022). "Business agility." Retrieved from: https://www.scaledagileframework.com/business-agility/

Schwab, K. (2016). "The Fourth Industrial Revolution: what it means, how to respond." World Economic Forum. Retrieved from: https://www.weforum.org/agenda/2016/01/the-fourth-industrial-revolution-what-it-means-and-how-to-respond/

Sibbet, D., and Drexler, A. (2018). The Drexler / Sibbet Team Performance Model®. Retrieved from https://www.thegrove.com/.

Solis, B. (2015). "The Engagement Gap: Why executives and employees think differently about employee engagement." Retrieved from: https://www.linkedin.com/pulse/engagement-gap-why-executives-employees-think-employee-brian-solis/

Stiglitz, J. (2020). "Here's how companies can build a better economy." Retrieved from: https://www.weforum.org/agenda/2020/02/davos-outcomes-economic-changes-progress/

Sull, D. (2009). "Competing through organizational agility." Retrieved from: https://www.mckinsey.com/business-functions/people-and-organizational-performance/our-insights/competing-through-organizational-agility

Taylor, S. (2016). "The 4 C's of generative leadership." Retrieved from: https://www.whatisdialogue.com/4-cs-generative-leadership/

Weiss, L. (2018). *How we work: Live your purpose, reclaiming your sanity and embrace the daily grind.* New York, NY: Harper Collins, LLC.

Westrum, R. (2004). "A typology of organisational cultures." Retrieved from: http://dx.doi.org/10.1136/qshc.2003.009522

White House (2022). "FACT SHEET: President Biden Calls for a Three-Month Federal Gas Tax Holiday."

Retrieved from: https://www.whitehouse.gov/briefing-room/statements-releases/2022/06/22/fact-sheet-president-biden-calls-for-a-three-month-federal-gas-tax-holiday/

What Did You Think of *Deliver Value: Happy Contributing People, Satisfied Customers, and Thriving Business?*

First, thank you for purchasing this book, **Deliver Value: Happy Contributing People, Satisfied Customers, and Thriving Business**. *I know you could have picked any number of books to read, but you picked this book, and for that, I am extremely grateful.*

I hope that it added value and quality to your everyday life. If so, it would be really nice if you could share this book with your friends and family by posting to Facebook.

If you enjoyed this book and found some benefit in reading this, I'd like to hear from you and hope that you could take some time to post a review on Amazon. Your feedback and support will help me to greatly improve my writing craft for future projects and make this book even better.

You can follow this link to **Deliver Value: Happy Contributing People, Satisfied Customers, and Thriving Business** *now.*

I want you, the reader, to know that your review is very important, so if you'd like to **leave a review**, *all you have to do is click* **here** *and away you go. I wish you all the best in your future success!*

Yours Truly, Dr. Dave

www.ingramcontent.com/pod-product-compliance
Lightning Source LLC
LaVergne TN
LVHW020233210926
848481LV00016B/1850